COOPERATIVE GAME THEORY AND APPLICATIONS

THEORY AND DECISION LIBRARY

General Editors: W. Leinfellner (*Vienna*) and G. Eberlein (*Munich*)

Series A: Philosophy and Methodology of the Social Sciences

Series B: Mathematical and Statistical Methods

Series C: Game Theory, Mathematical Programming and Operations Research

Series D: System Theory, Knowledge Engineering and Problem Solving

SERIES C: GAME THEORY, MATHEMATICAL PROGRAMMING AND OPERATIONS RESEARCH

VOLUME 16

Scope: Particular attention is paid in this series to game theory and operations research, their formal aspects and their applications to economic, political and social sciences as well as to socio-biology. It will encourage high standards in the application of game-theoretical methods to individual and social decision making.

The titles published in this series are listed at the end of this volume.

COOPERATIVE GAME THEORY AND APPLICATIONS

Cooperative Games Arising from Combinatorial Optimization Problems

by

IMMA CURIEL

University of Maryland Baltimore County

KLUWER ACADEMIC PUBLISHERS

BOSTON / DORDRECHT / LONDON

A C.I.P. Catalogue record for this book is available from the Library of Congress.

ISBN 978-1-4419-4775-8

Published by Kluwer Academic Publishers,
P.O. Box 17, 3300 AA Dordrecht, The Netherlands.

Kluwer Academic Publishers incorporates
the publishing programmes of
D. Reidel, Martinus Nijhoff, Dr W. Junk and MTP Press.

Sold and distributed in the U.S.A. and Canada
by Kluwer Academic Publishers,
101 Philip Drive, Norwell, MA 02061, U.S.A.

In all other countries, sold and distributed
by Kluwer Academic Publishers Group,
P.O. Box 322, 3300 AH Dordrecht, The Netherlands.

Printed on acid-free paper

Printed in the Netherlands

CONTENTS

PREFACE

After a period of extensive research in the theoretical foundations of Cooperative Game Theory, we are experiencing an increasing interest in methods for using the results that have been achieved. Applications are coming to the forefront of the research community's attention.

In this book I discuss applications of Cooperative Game Theory that arise from Combinatorial Optimization Problems. It is well known that the mathematical modelling of various real-world decision making situations gives rise to Combinatorial Optimization Problems. For situations where more than one decision maker is involved, classical Combinatorial Optimization Theory does not suffice, and it is here that Cooperative Game Theory can make an important contribution. If a group of decision makers decide to undertake a project together in order to increase the total revenue or decrease the total costs, they face two problems. The first one is how to execute the project in an optimal way. The second one is how to allocate the revenue or costs among the participants. It is quite easy to find examples where cooperation among several participants would have increased total revenue or decreased total costs, but that didn't work out because a reasonable allocation was not achieved. It is with this second problem that Cooperative Game Theory can help. The solution concepts from Cooperative Game Theory can be applied to arrive at allocation methods.

Although the choice of topics in this book is application driven I will also pay ample attention to theoretical questions that arise from the situations that are studied.

This book can be read in several ways. A reader who is not familiar with cooperative game theory can browse through it, looking especially at the examples in each chapter to get an idea of the concepts that play a role in that chapter. Since most concepts are illustrated with examples, in this way one can obtain a fairly good idea of what is happening. Somebody who wants to study the material in more, but not excruciating, detail can do this by considering the formal definitions and proofs of the parts that are of particular interest to her/him. Finally, of course, one can do a thorough reading of the book. In any case, I hope that the book will arise a genuine curiosity in the reader. That it will prompt him/her to pose questions and try to answer them, both about the material in this book as well as about other situations that may be modelled with Cooperative Game Theory.

This book consists of seven chapters. Each chapter starts with an example to illustrate the type of problems that I consider in that chapter. The book is self contained. In the first chapter the reader can find the definitions of concepts that are used throughout the book. Definitions that are used in one chapter only are given in that chapter. The second chapter discusses Linear Programming Games and extensions. Linear Programming Games provide a good stepping stone for Games based on Combinatorial Optimization Problems, just as Linear Programming provides a good stepping stone for Combinatorial Optimization. In the third chapter I treat Assignment Games and Permutation games together with their multidimensional generalizations. Economies with Indivisibilities and Matching Situations are also considered as models that are related to these games. Chapter four is dedicated to Sequencing Games and extensions. The topics in chapter five are Travelling Salesman Games and Routing Games. Chapter six deals with Minimum Cost Spanning Tree Games, and chapter seven concludes with Location Games.

Acknowledgements go to the Department of Mathematics and Statistics of the University of Maryland Baltimore County for providing me with a sabbatical year during which the greater part of this book was written, and to the Department of Social-Economic Studies of the University of The Netherlands Antilles for providing me with the space and equipment to work. Special thanks go to Johnny Djaoen of CCUNA, the Computer Centre of the University of the Netherlands Antilles, for his great way of giving help whenever I needed it. Very special thanks go to Robert Libier of the School of Engineering of the University of the Netherlands Antilles for his overall support.

Curaçao, August 1996   Imma Curiel

1

COOPERATIVE GAMES
AND SOLUTION CONCEPTS

1.1 INTRODUCTION

Let us consider the following situation. Five people, Anna, Bert, Carol, Dave, and Ellen decide to combine forces to start their own business. Each one of them has some skills and some capital to contribute to the venture. After careful analysis of the situation they conclude that they can achieve a yearly profit of 100 (in $10,000). Of course, this amount has to be divided among them. At first assigning 20 to each person seems a reasonable allocation. However, after some additional analysis Dave and Ellen figure out that if the two of them work together without the other three, they can make a yearly profit of 45. This is more than the 40 that they would receive according to the allocation above. Anna, Bert, and Carol also perform some additional analysis and realize that the three of them together can only make a profit of 25. So it is in their interest to keep Dave and Ellen in the group. Consequently, they decide to give 46 to Dave and Ellen, and to divide the remaining 54 equally among the three of them. This seems to settle the problem. To make sure that they cannot do better, Carol, Dave and Ellen decide to find out how much profit the three of them can make without the other two. It turns out that they can make a profit of 70, which is more than the 64 (46+18) that the second allocation described above gives to them. Anna and Bert do not have enough capital to start on their own, so they cannot make any profit with only the two of them. Considering this, they decide to give 71 to Carol, Dave and Ellen, and to divide the remaining 29 equally between the two of them. If Carol, Dave, and Ellen also decide to divide the 71 equally among the three of them some further analysis reveals the following problem. It turns out namely, that Bert, Dave, and Ellen can make a profit of 65 which is more than the last allocation assigns to them ($65 > 2 \times \frac{71}{3} + \frac{29}{2}$).

1

Meanwhile time is elapsing and they are losing precious business opportunities since none of them wants to commit without knowing first how the profit will be divided. They need a more systematic way to analyze the situation. The theory of cooperative games in characteristic function form yields a tool to do this. In modelling the situation as a cooperative game we consider the profit that each subgroup or coalition can achieve. Solution concepts in cooperative game theory describe profit allocations that take the profit of all these coalitions into consideration. In the following sections we will formally introduce cooperative games, we will discuss some properties of cooperative games, and we will study several solution concepts for cooperative games.

1.2 COOPERATIVE GAMES IN CHARACTERISTIC FUNCTION FORM

The following is the formal definition of a cooperative game in characteristic function form.

Definition 1.2.1 *A cooperative game in characteristic function form is an ordered pair $< N, v >$ where N is a finite set, the set of players, and the characteristic function v is a function from 2^N to R with $v(\emptyset) = 0$.*

A subset S of N is called a *coalition*. The number $v(S)$ can be regarded as the worth of coalition S in the game v. We will identify the game $< N, v >$ with the function v whenever there can be no confusion about the player set N. The set of all cooperative games with player set N will be denoted by G^N. Usually we will take the set N to be equal to $\{1, 2, \ldots, n\}$. For a coalition $S = \{i_1, i_2, \ldots, i_k\}$, $v(S)$ will be denoted by $v(i_1, i_2, \ldots, i_k)$ instead of $v(\{i_1, i_2, \ldots, i_k\})$.

Returning to the example of the introduction we see that there $N = \{$Anna, Bert, Carol, Dave, Ellen$\}$. In the table 1.1 the complete game is given. For convenience's sake we use Anna=1, Bert=2, Carol=3, Dave=4, and Ellen=5.

This table describes the profit that each of the coalitions can make. By examining the table carefully we see that when two coalitions with an empty intersection join, the profit of the new coalition is always at least equal to the sum of the profits of the constituting coalitions. Therefore we call this game superadditive. Formally,

S	v(S)	S	v(S)	S	v(S)	S	v(S)
{1}	0	{1,5}	20	{1,2,4}	35	{3,4,5}	70
{2}	0	{2,3}	15	{1,2,5}	40	{1,2,3,4}	60
{3}	0	{2,4}	25	{1,3,4}	40	{1,2,3,5}	65
{4}	5	{2,5}	30	{1,3,5}	45	{1,2,4,5}	75
{5}	10	{3,4}	30	{1,4,5}	55	{1,3,4,5}	80
{1,2}	0	{3,5}	35	{2,3,4}	50	{2,3,4,5}	90
{1,3}	5	{4,5}	45	{2,3,5}	55	{1,2,3,4,5}	100
{1,4}	15	{1,2,3}	25	{2,4,5}	65	∅	0

Table 1.1 The example of the introduction modelled as a cooperative game.

Definition 1.2.2 *A cooperative game v is said to be superadditive if*

$$v(S) + v(T) \leq v(S \cup T) \ \text{for all } S, T \in 2^N \ \text{with } S \cap T = \emptyset.$$

The game v is called subadditive if the reverse inequality holds and additive if equality holds.

If the game is superadditive one can say that the whole is more than the sum of the parts.
Further inspection of table 1.1. will show that an even stronger property holds. Namely, if two coalitions (not necessarily with an empty intersection) join, then the sum of the profits of the union and the intersection of these two, is at least equal to the sum of the profits of the constituting coalitions. In such a case the game is called convex. Formally,

Definition 1.2.3 *A cooperative game v is said to be convex if*

$$v(S) + v(T) \leq v(S \cup T) + v(S \cap T) \ \text{for all } S, T \in 2^N.$$

The game v is called concave if the reverse inequality holds.

Definition 1.2.4 *A cooperative game is called 0-normalized if $v(i) = 0$ for all $i \in N$, v is called (0,1)-normalized if $v(i) = 0$ for all $i \in N$ and $v(N) = 1$.*

Definition 1.2.5 *Two cooperative games u and v are said to be strategically equivalent if there exist a $k > 0$ and $\alpha_1, \alpha_2, \ldots, \alpha_n$ such that*

$$v(S) = ku(S) + \alpha(S) \ \text{for all } S \in 2^N.$$

Here and everywhere else we use $x(S)$ to denote $\sum_{i \in S} x_i$ for any $x \in R^n$, $S \subset N$.

Every cooperative game v with $v(N) > \sum_{i \in N} v(i)$ is strategic equivalent to a unique (0,1)-normalized game.

The set G^N of all cooperative games with player set N is a $(2^n - 1)$-dimensional linear space.

Definition 1.2.6 *For each $T \in 2^N$ the unanimity game u_T is defined by*

$$u_T(S) := \left\{ \begin{array}{ll} 1 & \text{if } S \subset T \\ 0 & \text{otherwise.} \end{array} \right.$$

The set $\{u_T | T \in 2^N \setminus \{\emptyset\}\}$ is a basis of the linear space G^N. For any game $v \in G^N$ we have

$$v = \sum_{\emptyset \neq T \subset N} c_T(v) u_T,$$

where

$$c_T(v) = \sum_{R \subset T} (-1)^{|T| - |R|} v(R).$$

The concept of *balancedness* is very important in the theory of cooperative games.

Definition 1.2.7 *A collection $\mathcal{B}$ of non-empty subsets of N is called a balanced collection if there exists positive numbers λ_S for all $S \in \mathcal{B}$ such that $\sum_{S \in \mathcal{B}} \lambda_S 1_S = 1_N$.*

The numbers λ_S are called the *weights* of the elements of $\mathcal{B}$. A balanced collection containing no proper balanced subcollection is called a *minimal balanced collection*.

Definition 1.2.8 *A cooperative game v is called a balanced game if for every balanced collection $\mathcal{B}$ with weights $\{\lambda_S\}_{S \in \mathcal{B}}$ the following holds:*

$$v(N) \geq \sum_{S \in \mathcal{B}} \lambda_S v(S).$$

Let v be a cooperative game. For each $S \in 2^N$ the *subgame* $< S, v_S >$ is defined by $v_S(T) = v(T)$ for all $T \subset S$.

Definition 1.2.9 *A cooperative game v is called totally balanced if for each coalition S the subgame v_S is balanced.*

1.3 THE CORE AND THE WEBER SET

As we saw in the example in the introduction, if all players in a game decide to work together then the question arises of how to divide the total profit. If one or more players conceive of a proposed allocation as being disadvantageous to them, they can decide to leave. This can happen even when the game is superadditive, i.e., when the maximum profit can be made by all players cooperating. We will use a payoff vector $x \in R^n$ to represent a distribution of $v(N)$ where $N = \{1, 2, \ldots, n\}$. Here x_i denotes the amount assigned to player $i \in N$. An $x \in R^n$ with $x(N) = v(N)$ is called *efficient*.

Definition 1.3.1 *A function f which assigns to every n-person cooperative game v a, possibly empty, subset f(v) of R^n is called a solution concept.*

A solution concept f is a one-point solution concept if $|f(v)| = 1$ for every v. If f is a one-point solution concept then $f(v)$ is denoted as an element of R^n rather than as a subset of R^n, i.e. $f(v) = x$ instead of $f(v) = \{x\}$.

Definition 1.3.2 *The set of pre-imputations $PI(v)$ of the game v is the subset of R^n that contains all efficient vectors for v.*

It is to be expected that, in evaluating a proposed allocation of total profits, each player will compare how much he obtains according to the allocation with how much he can make by working alone. If the comparison turns out to be unfavourable he will prefer to work alone. If this is not the case the payoff vector is called individual rational. Formally, a vector $x \in R^n$ is said to be *individual rational* for a game v if $x_i \geq v(i)$ for all $i \in N$.

Definition 1.3.3 *The imputation set $I(v)$ of a game v is defined by*

$$I(v) := \{x \in R^n | x(N) = v(N) \text{ and } x_i \geq v(i) \text{ for all } i \in N\}.$$

As we saw in the example in the introduction, even when an allocation is individual rational there may be a problem if a group of players figure out that they can do better by working without the others. This is not the case when x belongs to the *core* of v.

Definition 1.3.4 *The core $C(v)$ of v is defined by*

$$C(v) := \{x \in R^n | x(N) = v(N), \ x(S) \geq v(S) \text{ for all } S \in 2^N\}.$$

None of the allocations that we considered in the example in the introduction were in the core. Consequently, there was always a coalition that could do better by separating from the grand coalition.

The core of a game is a convex, compact subset of R^n and can be empty. A characterization of games with a non-empty core is given by the following theorem.

Theorem 1.3.5 (Bondareva, Shapley) *A cooperative game has a non-empty core if and only if it is balanced.*

The core of a game can be empty. The *Weber set* $W(v)$ of a game v is always non-empty and contains the core, cf. [140].

Definition 1.3.6 *Let π be a permutation of N. The marginal vector $m^\pi(v)$ of the game v is defined by*

$$m_i^\pi(v) := v(P(\pi, i) \cup \{i\}) - v(P(\pi, i))$$

where $P(\pi, i) := \{j \in N | \pi(j) < \pi(i)\}$ is the set of predecessors of i with respect to π.

Definition 1.3.7 *The Weber set $W(v)$ of v is the convex hull of the $n!$ marginal vectors.*

As stated above the Weber set always contains the core. For convex games the reverse is also true. For the 5-person game of table 1.1 there are 120 marginal vectors, but several of them coincide and we find that the core and the Weber set are equal to the convex hull of the vectors (0,0,25,35,40), (0,20,5,35,40), (0,20,25,15,40), (0,20,25,35,20), (10,0,15,35,40), (10,0,25,25,40), (10,0,25,35,30), (5,20,0,35,40), (10,15,0,35,40), (10,20,0,30,40), (10,20,0,35,35), (10,20,25,5,40), (10,20,25,35,10).

1.4 CONVEX GAMES AND THE MINIMARG AND MAXIMARG OPERATORS

Using the marginal vectors defined in the previous section we can define two operators on G^N. First an additive game m_v^π is defined as follows:

$$m_v^\pi(S) := \sum_{i \in S} m_i^\pi(v).$$

Definition 1.4.1 *The minimarg operator Mi assigns to each game $v \in G^N$ the game Mi(v) given by*

$$Mi(v)(S) := \min_{\pi \in \Pi_N} m_v^\pi(S) \text{ for all } S \in 2^N.$$

The maximarg operator Ma assigns to each game $v \in G^N$ the game Ma(v) given by

$$Ma(v)(S) := \max_{\pi \in \Pi_N} m_v^\pi(S) \text{ for all } S \in 2^N.$$

Here Π_N is the set of all permutations of N. The minimarg operator assigns to each game the minimum of the marginal values, while the maximarg operator assigns to each game the maximum of the marginal values. Assuming that the grand coalition forms according to some order, one can assign to each player his marginal contribution in this process. A coalition S can consider a worst possible order, i.e., an order for which the sum of the amounts that its members receive is minimal. The worth of S in this case is $Mi(v)(S)$. On the other hand, $Ma(v)(S)$ represents the best possible outcome for S.

Convex games can be characterized in different ways using marginal contributions.

Theorem 1.4.2 *The following five statements are equivalent.*

1. *The game v is convex.*

2. $v(S \cup \{i\}) - v(S) \geq v(T \cup \{i\}) - v(T)$ *for all $S, T \in 2^N$ with $T \subset S \subset N \setminus \{i\}$.*

3. $v(S \cup R) - v(S) \geq v(T \cup R) - v(T)$ *for all $S, T \in 2^N$ with $T \subset S \subset N \setminus R$.*

4. $W(v) = C(v)$.

S	$Ma(v)(S)$	S	$Ma(v)(S)$	S	$Ma(v)(S)$
{1}	10	{2,5}	60	{2,3,5}	85
{2}	20	{3,4}	60	{2,4,5}	95
{3}	25	{3,5}	65	{3,4,5}	100
{4}	35	{4,5}	75	{1,2,3,4}	90
{5}	40	{1,2,3}	55	{1,2,3,5}	95
{1,2}	30	{1,2,4}	65	{1,2,4,5}	100
{1,3}	35	{1,2,5}	70	{1,3,4,5}	100
{1,4}	45	{1,3,4}	70	{2,3,4,5}	100
{1,5}	50	{1,3,5}	75	{1,2,3,4,5}	100
{2,3}	45	{1,4,5}	85	∅	0
{2,4}	55	{2,3,4}	80		

Table 1.2   The game $Ma(v)$ for the game v of table 1.1.

5. $Mi(v) = v$.

The minimarg and the maximarg operators can be applied iteratively to a game. The following theorem was proved in [27].

Theorem 1.4.3 *Let $v \in G^N$. For $k = 1, 2, \ldots$, define the games v^k by $v^1 = v$ and $v^k = Mi(v^{k-1})$ ($v^k = Ma(v^{k-1})$). Then $\lim_{k \to \infty} v^k$ exists and is convex (concave).*

For several classes of games it has been shown that the limit will be reached after finitely many iterations. Cf. [27], [104].

Since the game of table 1.1 is convex applying the minimarg operator to it will yield the same game. The game $Ma(v)$ is given in table 1.2. This game is concave.

1.5   THE SHAPLEY-VALUE

Carol wants to investigate her position in the joint venture, that she and the others are considering, in some more detail to figure out how much she can expect to obtain from the total profit of 100. She calculates her marginal contribution in each possible order of formation of the grand coalition. If the grand coalition is formed in the order Anna, Dave, Ellen, Carol, Bert, for example, she would be joining the coalition {Anna, Dave, Ellen} which can make a profit of 55 without her, and a profit of 80 after she joins. So her marginal contribution is 25. This is how much she expects to obtain if the grand coalition is formed in this way. Since she has no reason to assume that one order of formation is more likely than any other, she assigns equal probability to each. Thus, she calculates her expected payoff to be $18\frac{1}{2}$. Each one of the others can make the same assessment. The payoff vector generated in this way is $(7\frac{1}{4}, 14\frac{3}{4}, 18\frac{1}{2}, 27\frac{1}{4}, 32\frac{1}{4})$. It is an element of the core of the game. It is the Shapley-value of the game. The *Shapley-value* ϕ assigns to the game v the average of the additive marginal games m_v^π. So

$$\phi(v) := \frac{1}{n!} \sum_{\pi \in \Pi_N} m_v^\pi$$

It follows that the Shapley-value of a convex game lies in the barycenter of the core of the game. The Shapley-value of a non-convex game need not be an element of the core of the game. A different way of looking at the Shapley-value is obtained by considering the axioms that Shapley used in defining it. In the following a one-point solution concept will also be called a solution function.

Definition 1.5.1 *A player in a game v is called a dummy player if $v(S \cup \{i\}) - v(S) = v(i)$ for all $S \subset N \setminus \{i\}$.*

Definition 1.5.2 *A solution function f is said to satisfy the dummy player property if $f_i(v) = v(i)$ whenever i is a dummy player in the game v.*

Definition 1.5.3 *A solution function f is said to be efficient if $f(v)$ is efficient for every game v.*

Definition 1.5.4 *A solution function f satisfies the symmetry property if $f_{\pi(i)}(\pi v) = f_i(v)$ for every game v, each player $i \in N$ and each permutation $\pi \in \Pi_N$. Here the game πv is defined by $\pi v(\pi(S)) := v(S)$ for all $S \in 2^N$.*

Definition 1.5.5 *A solution function f is said to be additive if $f(u + v) = f(u) + f(v)$ for all cooperative games u, v.*

The dummy player, efficiency, and additivity properties are straightforward to interpret. The symmetry property says that the payoff of a player should depend only on the role he plays in the game, not on who he is.

Shapley proved that there is a unique solution concept that satisfies these properties, the Shapley-value. An alternative formula for the Shapley-value is given below.

$$\phi_i(v) = \sum_{S \subset N \setminus \{i\}} \frac{|S|!(n-|S|)!}{n!} (v(S \cup \{i\}) - v(S)).$$

Several other axiomatic characterizations of the Shapley-value abound in the literature. Cf. [142], [34], [63].

1.6 THE BARGAINING SET

In discussing the allocation (20,20,20,20,20) in the example in the introduction, we saw that both Dave and Ellen could raise an objection against it since the two of them can make a profit of 45. The concept of an objection of a player is formalized and used in the definition of the bargaining set in [3]. Let

$$\Gamma_{ij} := \{S \in 2^N | i \in S, j \notin S\}.$$

An *objection* of player i against player j with respect to an imputation $x \in I(v)$ in the game $v \in G^N$ is a pair $(y; S)$ where $S \in \Gamma_{ij}$ and $y = (y_k)_{k \in S}$ is a $|S|$-tuple of real numbers satisfying

$$y(S) = v(S) \text{ and } y_k > x_k \text{ for all } k \in S.$$

A *counter objection* to the objection $(y; S)$ is a pair $(z; T)$ where $T \in \Gamma_{ji}$ and $z = (z_k)_{k \in T}$ is a $|T|$-tuple of real numbers such that

$$z(T) = v(T), \ z_k \geq y_k \text{ for } k \in S \cap T \text{ and } z_k \geq x_k \text{ for } k \in T \setminus S$$

Definition 1.6.1 *An imputation $x \in I(v)$ is said to belong to the bargaining set $\mathcal{M}(v)$ of the game v, if for any objection of one player against another with respect to x, there exists a counter objection.*

Since there is no objection possible with respect to an element of the core it follows that $C(v) \subset \mathcal{M}(v)$. The bargaining set is always non-empty. For convex games the bargaining set and the core coincide, cf. [81].

1.7 THE (PRE-)KERNEL AND THE (PRE-)NUCLEOLUS

To see how (un)happy a coalition S will be with a payoff vector x in a game v we can look at the *excess* $e(S, x)$ of S with respect to x defined by

$$e(S, x) := v(S) - x(S).$$

The smaller $e(S, x)$, the happier S will be with x. Note that $x \in C(v)$ if and only if $e(S, x) \leq 0$ for all $S \subset N$ and $e(N, x) = 0$.

If a payoff vector x has been proposed in the game v, player i can compare his position with that of player j by considering the *maximum surplus* $s_{ij}(x)$ *of* i *against* j with respect to x, defined by

$$s_{ij}(x) := \max_{S \in \Gamma_{ij}} e(S, x)$$

The maximum surplus of i against j with respect to x can be regarded as the highest payoff that player i can gain (or the minimal amount that i can lose if $s_{ij}(x)$ is negative) without the cooperation of j. Player i can do this by forming a coalition without j but with other players who are satisfied with their payoff according to x. Therefore, $s_{ij}(x)$ can be regarded as the weight of a possible threat of i against j. If x is an imputation then player j cannot be threatened by i or any other player when $x_j = v(j)$ since j can obtain $v(j)$ by operating alone. We say that i *outweighs* j if

$$x_j > v(j) \text{ and } s_{ij}(x) > s_{ji}(x).$$

The *kernel*, introduced in [28], consists of those imputations for which no player outweighs another one.

Definition 1.7.1 *The kernel* $\mathcal{K}(v)$ *of a game* v *is defined by*

$$\mathcal{K}(v) := \{x \in I(v) | s_{ij}(x) \leq s_{ji}(x) \text{ or } x_j = v(j) \text{ for all } i, j \in N\}.$$

Definition 1.7.2 *The pre-kernel* $\mathcal{PK}(v)$ *of a game* v *is defined by*

$$\mathcal{PK}(v) := \{x \in PI(v) | s_{ij}(x) = s_{ji}(x) \text{ for all } i, j \in N\}.$$

The kernel and the pre-kernel are always non-empty. The kernel is a subset of the bargaining set. For superadditive games the kernel and the pre-kernel coincide.

Since $e(S, x)$ is a measure of the (un)happiness of S with x we can try to find a payoff vector which minimizes the maximum excess. We construct a vector $\theta(x)$ by arranging the excesses of the 2^n subsets of N in decreasing order. So

$$\theta_i(x) \geq \theta_j(x) \text{ whenever } 1 \leq i \leq j \leq 2^n.$$

With $y <_L z$ we mean that y is lexicographically smaller than z while $y \leq_L z$ will be used to indicate that either $y <_L z$ or $y = z$.

Definition 1.7.3 *The nucleolus* $\nu(v)$ *of a game* v *is defined by*

$$\nu(v) := \{x \in I(v) | \theta(x) \leq_L \theta(y) \text{ for all } y \in I(v)\}.$$

The nucleolus of a game was introduced by Schmeidler [114] who also showed that it always consists of one point which is an element of the kernel and which is in the core whenever the core is non-empty. For a convex game v we have $\mathcal{PK}(v) = \mathcal{K}(v) = \nu(v)$.

Definition 1.7.4 *The pre-nucleolus* $p\nu(v)$ *of a game* v *is defined by*

$$p\nu(v) := \{x \in PI(v) | \theta(x) \leq_L \theta(y) \text{ for all } y \in PI(v)\}.$$

The pre-nucleolus always consists of one point.

For the game of table 1.1 we have $\mathcal{PK}(v) = \mathcal{K}(v) = \nu(v) = (5, 13\frac{3}{4}, 18\frac{3}{4}, 28\frac{3}{4}, 33\frac{3}{4})$.

1.8   THE τ-VALUE

Dave decides to figure out how much the maximum amount is, that he can expect to receive from the total profit of 100. He thinks: "The others will never give me more than the difference between the profit we make when I join and the profit they can make without me. So the most I can expect is 100-65=35." He also wants to decide how much the minimum amount is that he will accept. He reasons as follows: "Each one of the others can compute the maximum amount that they can expect to get in the same way as I did. If I am in a coalition with some of them then they will be happy to receive this maximum amount and leave the remainder of the profit that this coalition can make to me. So I should find a coalition that maximizes this remainder. In this way I will get a lower bound on what I will accept." After some computations he realizes that all coalitions yield the same value of 5. Now he has an upper bound of 35 and a lower bound of 5. His real payoff will be in between these two. The others can reason in the same way. Anna's upper and lower bound are 10 and 0, Bert's are 20 and 0, Carol's are 25 and 0, and Ellen's are 40 and 10. The vector of upper bounds is (10,20,25,35,40) and the vector of lower bounds is (0,0,0,5,10). The payoff vector that is a convex combination of these two and satisfies the efficiency property is $(7\frac{9}{23}, 14\frac{18}{23}, 18\frac{11}{23}, 27\frac{4}{23}, 32\frac{4}{23})$. This is the τ-*value* of the game. Let's formalize all this.

Definition 1.8.1 *The upper vector* M^v *of the game* v *is defined by*

$$M_i^v := v(N) - v(N \setminus \{i\}) \text{ for all } i \in N.$$

The lower vector μ^v *of* v *is defined by*

$$\mu_i^v := \max_{S:S \ni i} (v(S) - \sum_{j \in S \setminus \{i\}} M_j^v)$$

Definition 1.8.2 *A game* $v \in G^N$ *is called quasi-balanced if*

1. $\mu^v \leq M^v$

2. $\mu^v(N) \leq v(N) \leq M^v(N)$.

Every balanced game is quasi-balanced. The τ-value of a quasi-balanced game was introduced in [131].

Definition 1.8.3 *The* τ-*value of a quasi-balanced game* v *is defined by*

$$\tau(v) := \lambda \mu^v + (1 - \lambda) M^v$$

where $\lambda \in [0,1]$ *is uniquely determined by* $\sum_{i \in N} \tau_i(v) = v(N)$.

The τ-value of v need not be an element of the core, even when this is non-empty. Driessen and Tijs [38] give necessary and sufficient conditions for $\tau(v)$ to be an element of the core.

1.9 SEMICONVEX AND 1-CONVEX GAMES

In general the τ-value is cumbersome to compute because in computing the vector μ_i^v we have to consider all the coalitions containing i. For the two classes of games studied in this section, the τ-value can be computed with a lot less work.

Definition 1.9.1 *The gap function $g^v : 2^N \to R$ of the game $v \in G^N$ is defined by*
$$g^v(S) := M^v(S) - v(S) \text{ for all } S \in 2^N.$$

Definition 1.9.2 *A game $v \in G^N$ is called semiconvex if v is superadditive and $g^v(i) \leq g^v(S)$ for all $i \in N$ and $S \in 2^N$ with $i \in S$. If the reverse inequalities hold and v is subadditive, then v is called semiconcave.*

Every convex game is semiconvex and every semiconvex game is quasi-balanced.

Theorem 1.9.3 (Driessen and Tijs) *Let v be a semiconvex game. Then $\tau(v) = \lambda \underline{v} + (1 - \lambda)M^v$ where $\underline{v} = (v(1), v(2), \ldots, v(n))$ and where λ is such that $\tau(v)$ is efficient.*

Definition 1.9.4 *A game $v \in G^N$ is called 1-convex if*
$$0 \leq g^v(N) \leq g^v(S) \text{ for all } S \subset N, \, S \neq \emptyset.$$

If the reverse inequalities hold then v is called 1-concave.

Theorem 1.9.5 (Driessen and Tijs) *Let v be a 1-convex game. Then $\tau(v)$ is given by*
$$\tau_i(v) = M_i^v - \frac{1}{n}g^v(N).$$

Theorem 1.9.6 (Driessen) *The game v is a 1-convex game if and only if the extreme points of $C(v)$ are the vectors $M^v - g^v(N)f^i$ for all $i \in N$, where the vector f^i is the vector with 1 in the i-th position and 0 everywhere else.*

We see that for a 1-convex game the τ-value is the barycenter of the core.

Theorem 1.9.7 (Driessen and Tijs) *Let v be a 1-convex game. Then* $\nu(v) = \tau(v)$.

So for 1-convex games not only the τ-value but also the nucleolus is easy to compute.

1.10   SIMPLE GAMES

Simple games are very well suited for modelling situations of voting and committee control. It is for the latter that we will be using them.

Definition 1.10.1 *A cooperative game $v \in G^N$ is called a simple game if $v(N) = 1$ and $v(S) \in \{0,1\}$ for all $S \in 2^N$.*

Definition 1.10.2 *A simple game v is called monotonic if $v(S) \le v(T)$ for all coalitions S and T with $S \subset T$.*

Definition 1.10.3 *A coalition S in a simple game is called winning if $v(S) = 1$, S is called losing if $v(S) = 0$.*

Definition 1.10.4 *A coalition S in a monotonic simple game v is called minimal winning if $v(S) = 1$ and $v(T) = 0$ for all $T \subset S$, $T \ne S$.*

Definition 1.10.5 *A player $i \in N$ in a simple game v is said to be:*

a dummy player	*if*	$v(S) + v(i) = v(S \cup \{i\})$ *for all $S \subset N \setminus \{i\}$*
a null player	*if*	$v(S) = v(S \cup \{i\})$ *for all $S \subset N \setminus \{i\}$*
a dictator	*if*	$v(S)=1$ *if and only if $i \in S$*
a veto player	*if*	$v(S) = 1$ *implies $i \in S$.*

A dummy player i can be either a null player, in case $v(i) = 0$, or a dictator in case $v(i) = 1$.

Theorem 1.10.6 *A simple game v is balanced if and only if v has veto players.*

It is easy to see that any non-negative division of 1 among the veto players will be an element of the core.

Spinetto [125] proved that every non-negative (0,1)-normalized balanced game can be written as a convex combination of balanced simple games. Derks [31] generalized this result to show that every non-negative balanced game can be written as a positive linear combination of simple games.

As already mentioned simple games, and especially monotonic simple games, can be used for modelling situations of committee control and voting. A player in such a situation will want to know how much influence he has on the outcome of the game. He will need a measure of his power in the game. Such a measure is called a *power index*.

Definition 1.10.7 *A power index is a function f which assigns to every monotonic simple game $v \in G^N$ a vector $f(v) \in R^n$.*

Two well-known power indices are the Shapley-Shubik index and the Banzhaf index. The Shapley-Shubik index is in fact the restriction of the Shapley-value to the monotonic simple games. The Banzhaf index counts the number of *swings* of a player i. A swing for player i is a pair of coalitions $(S \cup \{i\}, S \setminus \{i\})$ such that $S \cup \{i\}$ is winning and $S \setminus \{i\}$ is losing.

In [16], [29], [30], [96] power indices, that depend only on the minimal winning coalitions of a monotonic simple game, are discussed.

1.11   COST GAMES

Anna, Bert, Carol, Dave, and Ellen each want to make some improvements on their houses. Comparing how much it will cost each of them if they work separately with how much it will cost if together they bring in a contractor, they realize that they can save on the total costs by cooperating. Of course, they will have to agree on how much of the total costs each of them will pay. In fact, they face the same type of problem as before, only now with costs instead of profit. Whereas everybody would like to get as much as possible of the profit, everybody would like to pay as little as possible of the costs. This situation can be modelled using a *cooperative cost game*. Table 1.3 describes this game. All the solution concepts that have been introduced in the previous sections can be adapted to be used for cost games. This should be done while

S	$c(S)$	S	$c(S)$	S	$c(S)$	S	$c(S)$
{1}	10	{1,5}	50	{1,2,4}	65	{3,4,5}	100
{2}	20	{2,3}	45	{1,2,5}	70	{1,2,3,4}	90
{3}	25	{2,4}	55	{1,3,4}	70	{1,2,3,5}	95
{4}	35	{2,5}	60	{1,3,5}	75	{1,2,4,5}	100
{5}	40	{3,4}	60	{1,4,5}	85	{1,3,4,5}	100
{1,2}	30	{3,5}	65	{2,3,4}	80	{2,3,4,5}	100
{1,3}	35	{4,5}	75	{2,3,5}	85	{1,2,3,4,5}	100
{1,4}	45	{1,2,3}	55	{2,4,5}	95	∅	0

Table 1.3   A 5-person cost game.

keeping the intuitive explanation given for each solution concept in mind. The reader is invited to do this, and to check that the game given in table 1.3 is (semi)concave, and that it has the same core, Shapley-value, bargaining set, (pre-)kernel, nucleolus, and τ-value as the game from table 1.1. This is due to the fact that v is (semi)convex and that c is the *dual game*[1] of v.

---

[1]The dual game of a game v is the game v^* defined by $v^*(S) = v(N) - v(N \setminus S)$ for all $S \in 2^N$.

LINEAR PROGRAMMING GAMES

2.1 INTRODUCTION

Three friends, Fred, Grace, and Henry, are looking into the possibilities of starting a day care center for children from (practically) 0 to 4 years. After a thorough investigation of the local regulations and market, they come to the following assessment of the situation. They need one nursemaid per four children from 0 to 2, and one nursemaid per 10 children from 2 to 4. For each child from 0 to 2 they are required to have 8 square meters available inside, and 4 outside. For a child from 2 to 4 the requirements are 5 and 6, respectively. After calculating all costs they realize that they can make a net profit of $200 per month per child from 0 to 2 and a net profit of $150 per month per child from 2 to 4. (The difference is due to the fact that since there are not many day care centers that accept children form 0 to 2 years they can ask a higher fee for them.) Fred knows 9 people whom he can hire as nursemaids. He has the possibility of renting 260 square meters inside and 200 outside. While considering the possibility of starting a day care center on his own he wants to calculate the maximum profit that he can make while complying with all regulations. Therefore he has to solve the following linear programming problem.[1]

$$\begin{array}{llll}
\max & 200x_1 & + & 150x_2 \\
\text{s.t.} & 0.25x_1 & + & 0.10x_2 & \leq & 9 & \text{(nursemaid requirement)} \\
& 8x_1 & + & 5x_2 & \leq & 260 & \text{(inside space requirement)} \\
& 4x_1 & + & 6x_2 & \leq & 200 & \text{(outside space requirement)} \\
& x_1 & \geq & 0,\, x_2 & \geq & 0
\end{array}$$

Here x_1 denotes the number of children from 0 to 2 that he should accept and x_2 denotes he number of children form 2 to 4 that he should accept. By solving this LP-problem he sees that the best thing for him to do is to accept 20 children from each age class which will give him a profit of $7000.

Grace knows 5 people whom she can hire as nursemaids. She has the possibility of renting 120 square meters inside and 200 outside. To maximize her profit she needs to solve the following LP-problem.

$$\begin{array}{llll}
\max & 200x_1 & + & 150x_2 \\
\text{s.t.} & 0.25x_1 & + & 0.10x_2 & \leq & 5 \\
& 8x_1 & + & 5x_2 & \leq & 120 \\
& 4x_1 & + & 6x_2 & \leq & 200 \\
& x_1 & \geq & 0,\, x_2 & \geq & 0
\end{array}$$

Solving this reveals that she should accept no children from 0 to 2 and 24 from 2 to 4, y elding a profit of $3600.

Henry knows 14 people whom he can hire as nursemaids. He has the possibility of renting 590 square meters inside and 400 outside. He faces the following LP-problem.

$$\begin{array}{llll}
\max & 200x_1 & + & 150x_2 \\
\text{s.t.} & 0.25x_1 & + & 0.10x_2 & \leq & 14 \\
& 8x_1 & + & 5x_2 & \leq & 590 \\
& 4x_1 & + & 6x_2 & \leq & 400 \\
& x_1 & \geq & 0,\, x_2 & \geq & 0
\end{array}$$

The best thing for him to do is to accept 40 of each age group, which will give him a profit of $14,000

They can also decide to combine forces. We assume that there is no overlap between the people that they can hire as nursemaids. If Fred and Grace decide

---

[1]In fact, Fred has to solve an integer programming problem and not a linear programming problem since children are indivisible. As all the numbers in this example have been chosen carefully to ensure that the optimal solutions of all LP-problems that occur are integers, we will sweep this complication under the carpet.

S	$v(S)$	S	$v(S)$	S	$v(S)$	S	$v(S)$
{1}	7,000	{3}	14,000	{1,3}	22,500	{1,2,3}	26,500
{2}	3,600	{1,2}	11,000	{2,3}	19,500	∅	0

Table 2.1 The day care center game.

to work together they have to solve the following LP-problem.

$$
\begin{aligned}
\max \quad & 200x_1 + 150x_2 \\
\text{s.t.} \quad & 0.25x_1 + 0.10x_2 \leq 14 \\
& 8x_1 + 5x_2 \leq 380 \\
& 4x_1 + 6x_2 \leq 400 \\
& x_1 \geq 0, \; x_2 \geq 0
\end{aligned}
$$

Solving this reveals that they should accept 10 children from 0 to 2 and 60 from 2 to 4, yielding a profit of \$11,000. Similarly, one can compute the best thing to do if Fred and Henry cooperate, if Grace and Henry cooperate, and if they all three cooperate. Fred and Henry should accept 75 children from 0 to 2 and 50 from 2 to 4. This will give them a profit of \$22,500. Grace and Henry should accept 45 children from 0 to 2 and 70 from 2 to 4. This will give them a profit of 19,500. If all three of them work together they have to solve the following LP-problem.

$$
\begin{aligned}
\max \quad & 200x_1 + 150x_2 \\
\text{s.t.} \quad & 0.25x_1 + 0.10x_2 \leq 28 \\
& 8x_1 + 5x_2 \leq 970 \\
& 4x_1 + 6x_2 \leq 800 \\
& x_1 \geq 0, \; x_2 \geq 0
\end{aligned}
$$

They should accept 65 children from 0 to 2 and 90 from 2 to 4, yielding a profit of \$26,500. Denoting Fred=1, Grace=2, Henry=3, this situation is described in table 2.1 as a cooperative game. Their total profit is maximized if they combine forces. In this case they will have to figure out a way to divide the profit of \$26,500. One way of doing this is to consider the worth of the assets that each one of them brings with him/her. To do this we will consider the dual problem to the LP-problem that determines $v(N)$.

$$
\begin{aligned}
\min \quad & 28y_1 + 970y_2 + 800y_3 \\
\text{s.t.} \quad & 0.25y_1 + 8y_2 + 4y_3 \geq 200 \\
& 0.10y_1 + 5y_2 + 6y_3 \geq 150 \\
& y_1 \geq 0, \; y_2 \geq 0, \; y_3 \geq 0
\end{aligned}
$$

The solution to this problem is $y_1 = 0$, $y_2 = 21\frac{3}{7}$, $y_3 = 7\frac{1}{7}$. These are considered to be shadow prices for the nursemaids, inside space, and outside space, respectively. A payoff vector can be constructed with them as follows. Fred receives

$$0 \times 9 + 21\frac{3}{7} \times 260 + 7\frac{1}{7} \times 200 = 7000.$$

Grace receives

$$0 \times 5 + 21\frac{3}{7} \times 120 + 7\frac{1}{7} \times 200 = 4000.$$

Henry receives

$$0 \times 14 + 21\frac{3}{7} \times 590 + 7\frac{1}{7} \times 400 = 15,500.$$

It is easy to check that this allocation is an element of the core of the game. In the next section we will see that this is not a coincidence. There we will formally introduce linear programming games and we will show that they are totally balanced. In fact, every totally balanced game is a linear programming game. We will also introduce and study flow games. In Section 2.3 we will look at linear programming games with committee control. The committee control will be modelled with the aid of simple games. The existence of veto players in these simple games is important for results concerning the cores of these games. In section 2.4 we will consider non-balanced linear programming games and we will discuss several ways to arrive at an allocation of the revenues in such a game. Section 2.5 deals with linear programming games in which the players stake claims on the resources. This chapter concludes with a section on simple flow games, in which the capacity of each arc is equal to 1, and each player owns exactly one arc.

2.2   LINEAR PROGRAMMING GAMES

Consider the following linear programming problem.

$$\begin{array}{rl} \max & c \cdot x \\ \text{s.t.} & xA \leq b \\ & xH = d \\ & x \geq 0 \end{array} \qquad (2.1)$$

Here $c \in R^m$, $b \in R^p$, $d \in R^r$, A is an $m \times p$-matrix, H is an $m \times r$-matrix. An $x \in R^m$ which satisfies the constraints is called a *solution* of this problem. If there exists a solution the problem is called *feasible*, otherwise it is called *infeasible*. A solution $\hat{x}$ of 2.1 is an *optimal solution* if $c \cdot \hat{x} \geq c \cdot x$ for all solutions x. In this case the *value* $v_p(A, H, b, d, c)$ of 2.1 is equal to $c \cdot \hat{x}$. The *dual* problem of 2.1 is

$$
\begin{aligned}
\min \quad & y \cdot b + z \cdot d \\
\text{s.t.} \quad & Ay + Hz \geq c \\
& y \geq 0
\end{aligned}
\tag{2.2}
$$

In an obvious way the definitions given above can be adapted to problem 2.2. The value of 2.2 is denoted by $v_d(A, H, b, d, c)$. From the duality theorem of linear programming it follows that 2.2 is feasible and bounded if and only if 2.1 is feasible and bounded, and then $v_p(A, H, b, d, c) = v_d(A, H, b, d, c)$.

A cooperative game can be constructed from 2.1 by making all, or some of the right hand sides in the constraints, depend on the coalitions. This can be done in several ways. In this section we study the following case. For every $i \in N$ and every $k \in \{1, \ldots, p\}$ a $b_k(i) \in R$ is given. Let $b_k(S) := \sum_{i \in S} b_k(i)$ for all $S \in 2^N \setminus \{\emptyset\}$, and let $b(S)$ be the vector in R^p with k-th coordinate equal to $b_k(S)$. Similarly vectors $d(S) \in R^r$ are constructed from given $d_k(i) \in R$.

Definition 2.2.1 *A cooperative game v is called a linear programming game if there exist an $m \times p$ matrix A, an $m \times r$ matrix H, and vectors $b(S) \in R^p$ and $d(S) \in R^r$ for all $S \in 2^N \setminus \{\emptyset\}$ such that $v(S) = v_p(A, H, b(S), d(S), c)$.*

In order to obtain a game $v_p(A, H, b(S), d(S), c)$ should be a real number for all $S \in 2^n \setminus \{\emptyset\}$. So the LP-problem with right hand sides equal to $b(S)$ and $d(S)$ should be feasible and bounded. Whenever the equality constraints are void, a sufficient but not necessary condition for this to hold is that all the entries in the matrix A are non-negative, with in each row at least one positive entry, and that $b(S)$ is non-negative. The day care center game of the introduction is an example of a linear programming game.

Theorem 2.2.2 *Linear programming games are totally balanced.*

Proof. Let v be a linear programming game. Then for every $T \in 2^N \setminus \{\emptyset\}$ the subgame v_T of v is also a linear programming game. Consequently, it is sufficient to prove that v is balanced. Let $(\hat{y}, \hat{z}) \in R^p \times R^r$ be an optimal solution for the dual problem of the problem which determines $v(N)$. Define $u \in R^n$ by

$$
u_i := \hat{y} \cdot b(i) + \hat{z} \cdot d(i).
$$

Then

$$
u(N) = \hat{y} \cdot \sum_{i \in N} b(i) + \hat{z} \cdot \sum_{i \in N} d(i)
$$

$$= \hat{y} \cdot b(N) + \hat{z} \cdot d(N) = v(N).$$

Further,

$$u(S) = \hat{y} \cdot b(S) + \hat{z} \cdot d(S) \geq v(S) \text{ for all } S \in 2^N \setminus \{\emptyset\}.$$

The last inequality follows from the fact that $(\hat{y}, \hat{z})$ is also a solution of the dual problem of the problem which determines $v(S)$. Thus, it follows that $u \in C(v)$.
□

The proof of theorem 2.2.2 describes a less laborious way of arriving at a core element of v, if v is an LP-game. It is not necessary to compute the worth of all $2^n - 1$ non-empty coalitions. Just by solving one LP-problem one obtains an element of $C(v)$.

Theorem 2.2.3 *Every totally balanced game is a linear programming game.*

Proof. Let v be a totally balanced game. For each $T \in 2^N \setminus \{\emptyset\}$ let z^T be defined by $z_i^T := M + 1$ for all $i \notin T$ and such that its projection on R^T is an element of $C(v_T)$. Here $M \geq v(S)$ for all $S \in 2^N$. Let $T_1, T_2, \ldots, T_{2^n-1}$ be an ordering of the non-empty coalitions. Let A be the $n \times 2^n - 1$-matrix with all entries equal to one. For each $i \in N$ define the vector $b(i) \in R^{2^n-1}$ by $b_k(i) := z_i^{T_k}$. Then, for every $S \in 2^N \setminus \{\emptyset\}$, the worth $v(S)$ of S is equal to the value of the linear programming problem

$$\begin{array}{ll} \max & c \cdot x \\ \text{s.t.} & xA \leq b(S) \end{array} \tag{2.3}$$

Here $c \in R^n$ is the vector with all coordinates equal to one and $b(S) = \sum_{i \in S} b(i)$. Let A' be the $2n \times 2^n - 1$-matrix with j-th row equal to the j-th row of A for $1 \leq j \leq n$ and equal to minus the $(n - j)$-th row of A for $n + 1 \leq j \leq 2n$. Let $c' \in R^{2n}$ be the vector with the first n coordinates equal to 1 and the last n coordinates equal to -1. Problem 2.3 is equivalent to

$$\begin{array}{ll} \max & c' \cdot x \\ \text{s.t.} & xA' \leq b(S) \\ & x \geq 0 \end{array}$$

Thus it follows that v is a linear programming game with $v(S) = v_p(A', H, b(S), d(S), c')$ for all $S \in 2^N \setminus \{\emptyset\}$, where H is the $2n \times 1$-matrix with all entries equal to 0, and $d(S) = 0$ for all $S \in 2^N \setminus \{\emptyset\}$.   □

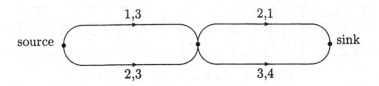

Figure 2.1   A graph leading to a flow game.

For certain classes of games there is a more natural way, than the one described in the proof of the previous theorem, to write them as LP-games. One such class is that of *flow games*. Let G be a directed graph with set of vertices P and set of arcs L. For every $p \in P$ let $B(p)$ denote the set of arcs which start in p and $E(p)$ the set of arcs which end in p. Two functions c and o are given on L. The function c is called the *capacity function* and assigns to every $l \in L$ a non-negative number, the capacity $c(l)$ of l. The function o is called the *ownership function* and assigns to every $l \in L$ a player, the owner $o(l)$ of l. We distinguish two different vertices from the others, a *source* s and a *sink* t. A *flow* from source to sink in such a graph is a function f from L to R with $0 \le f(l) \le c(l)$ and such that for every $p \in P \setminus \{s, t\}$ the amount of flow which enters p is equal to the amount which leaves p, i.e.,

$$\sum_{l \in B(p)} f(l) = \sum_{l \in E(p)} f(l). \tag{2.4}$$

The *value* of such a flow is the net amount which enters the sink and is equal to the net amount which leaves the source. A flow with a maximum value is called a maximum flow and this value is denoted by $w(G)$.

For each $S \in 2^N \setminus \{\emptyset\}$ a graph G_S can be obtained from G by keeping all vertices and removing all arcs which do not belong to a member of S. This new set of arcs is denoted by L_S.

Definition 2.2.4 *A game v is said to be a flow game if there exists a graph G with a capacity function c and an ownership function o such that $v(S) = w(G_S)$ for all $S \in 2^N \setminus \{\emptyset\}$.*

Example Let G be the graph given in figure 2.1. The first number at an arc indicates the owner, the second the capacity. The flow game arising from this graph is given in table 2.2. The core of this game is the convex hull of the points $(0,1,4), (0,2,3), (2,1,2), (2,2,1)$.

Theorem 2.2.5 *Flow games are linear programming games.*

S	$v(S)$	S	$v(S)$	S	$v(S)$	S	$v(S)$
{1}	0	{3}	0	{1,3}	3	{1,2,3}	5
{2}	1	{1,2}	1	{2,3}	3	$\emptyset$	0

Table 2.2 The flow game of figure 2.1.

Proof. Let v be a flow game with underlying graph G. Let $l_1, l_2, \ldots, l_r$ be an ordering of the arcs such that $\{l_1, l_2, \ldots, l_g\} = B(s)$ and $\{l_{g+1}, \ldots, l_h\} = E(s)$. For every $i \in N$ define the vector $b(i) \in R^r$ by $b_k(i) = c(l_k)$ if $o(l_k) = i$, $b_k(i) = 0$ otherwise. Let $s = p_1, p_2, \ldots, p_q = t$ be an ordering of the vertices. Define the $r \times q$-matrix $H = [h_{km}]$ to be almost the incidence matrix of the graph G. That is,

$$ h_{km} := \begin{cases} 1 & \text{if } m \neq 1, q \text{ and } l_k \in B(p_m) \\ -1 & \text{if } m \neq 1, q \text{ and } l_k \in E(p_m) \\ 0 & \text{otherwise.} \end{cases} $$

For every $i \in N$ let $d(i) \in R^q$ be the zero vector. Let $p \in R^r$ be the vector with the first g coordinates equal to 1, the subsequent $h - g$ coordinates equal to -1, and the remainder of the coordinates equal to 0. Then $v(S) = v_p(I, H, b(S), d(S), p)$. Here I is the $r \times r$-identity matrix. Thus, v is a linear programming game. The inequality constraints correspond to the capacity constraints, while the equality constraints correspond to the constraints 2.4. $\square$

It follows that flow games are totally balanced. Kalai and Zemel [66] proved this by using the Ford-Fulkerson max flow-min cut theorem which is a special case of the duality theorem of linear programming. In fact, Kalai and Zemel proved that every non-negative totally balanced game is a flow game.

A generalization of flow games is given by *flow games with losses*. Especially when considering the transportation of perishable goods it seems plausible to assume that there will be some difference between the amount at the beginning of an arc and the amount that arrives at the endpoint of the arc. Here we consider the case where the losses are proportional to the amount that is put through an arc. So, there exists a $\lambda \in [0, 1]$ such that if a flow $f(l)$ is put through arc l from p_1 to p_2 only $\lambda f(l)$ reaches p_2. In such a graph the equations 2.4 have to be replaced by

$$ \sum_{l \in B(p)} f(l) = \lambda \sum_{l \in E(p)} f(l) \quad \text{for all } p \in P \setminus \{s, t\}. $$

The value of such a flow is the net amount that reaches the sink. Note that, contrary to the case without losses, this will not be equal to the net amount that leaves the source. The value of a maximum flow in such a graph G is denoted by $w(G, \lambda)$.

Theorem 2.2.6 *Flow games with losses are linear programming games.*

Proof. Let v be a flow game with losses. Let $l_1, \ldots, l_r$ be an ordering of the arcs such that $\{l_{h+1}, \ldots, l_r\} = E(t)$ and $\{l_{g+1}, \ldots, l_h\} = B(t)$. For every $i \in N$ define the vector $b(i) \in R^r$ by $b_k(i) = c(l_k)$ if $o(l_k) = i$, $b_k(i) = 0$ otherwise. Let $s = p_1, p_2, \ldots, p_q = t$ be an ordering of the vertices. Define the $r \times q$-matrix $H = [h_{km}]$ by

$$
h_{km} = \begin{cases} 1 & \text{if } m \neq 1, q \text{ and } l_k \in B(p_m) \\ -\lambda & \text{if } m \neq 1, q \text{ and } l_k \in E(p_m) \\ 0 & \text{otherwise} \end{cases}
$$

For every $i \in N$ let $d(i)$ be the zero vector in R^q. Let $p \in R^r$ be the vector with the last $r - h$ coordinates equal to 1, the $h - g$ coordinates before those equal to -1, and the remainder of the coordinates equal to 0. Then $v(s) = v_p(I, H, b(S), d(S), p)$. Here I is the $r \times r$-identity matrix. $\qquad \square$

It follows that flow games with losses are totally balanced.

Another class of linear programming games is that of *linear production games* introduced by Owen [91]. In these games each player owns certain amounts of m different resources. The *resource vector* of player i is $b(i)$ with $b_k(i) \geq 0$ being the amount of the k-th resource that i possesses. These resources can be used to produce r different products. An $r \times q$-*production matrix* $A = [a_{jk}]$ gives the quantities needed of each resource to produce one unit of each product. This means that a_{jk} is the amount needed of the k-th resource to produce one unit of the j-th product. Each unit of product j can be sold at a given market price p_j. Every player wants to maximize his profit by producing a bundle of products that will yield most when sold. The players can pool their resources. In this way we obtain a linear production game with the values $v(S)$ given by

$$
\begin{aligned}
\max \quad & p \cdot x \\
\text{s.t.} \quad & xA \leq b(S) \\
& x \geq 0
\end{aligned} \tag{2.5}
$$

It is obvious that a linear production game is a linear programming game. The optimal solution $\hat{y}$ of the dual problem described in the proof of theorem 2.2.1

can be interpreted as shadow prices for the corresponding resources. The core element constructed from $\hat{y}$ can then be viewed as a way of paying to every player the amount that he would get, if instead of producing anything he would sell his resources.

As a generalization of linear production games we will discuss linear production games with transfer possibilities which were studied by Feltkamp et al. [45]. In these games it is possible to make products at several production sites. Transfers of products, resources, and technology between the production sites is possible. An example will clarify this.

Example Inez, John and Karin have facilities to produce six products at three different locations. The same two resources are needed for all these six products. In Freetown they can produce product p_1 and product p_2. In Garret they can produce product p_3 and product p_4. In Hyattsville they can produce product p_5 and product p_6. The production-matrices A^F, A^G, and A^H are given below.

$$A^F = \begin{pmatrix} 1 & 2 \\ 3 & 1 \end{pmatrix} A^G = \begin{pmatrix} 2 & 0 \\ 1 & 2 \end{pmatrix} A^H = \begin{pmatrix} 0 & 1 \\ 1 & 1 \end{pmatrix}$$

The vectors of resources of Inez, John and Karin at the several locations are given below. Here Inez=1, John=2, and Karin=3.

$$\begin{array}{lll} b^F(1) = (4,0) & b^G(1) = (2,3) & b^H(1) = (1,1) \\ b^F(2) = (2,2) & b^G(2) = (0,4) & b^H(2) = (3,0) \\ b^F(3) = (1,2) & b^G(3) = (2,1) & b^H(3) = (0,2) \end{array}$$

The price vectors at the three locations are given below. It is possible to sell all six products at each location.

$$p^F = (1,2,3,1,2,1) \, p^G = (2,1,1,1,2,1) \, p^H = (3,1,2,3,2,1)$$

If we consider the situation without transfers then each coalition faces three separate linear programming problems at the three different sites to maximize its profits. For the coalition consisting of only Inez these are:

$$\begin{array}{rrrrcl}
\max & x_1 & + & 2x_2 & & \\
\text{s.t.} & x_1 & + & 3x_2 & \leq & 4 \\
 & 2x_1 & + & x_2 & \leq & 0 \\
 & x_1 & \geq & 0, x_2 & \geq & 0
\end{array}
\qquad
\begin{array}{rrrrcl}
\max & x_3 & + & x_4 & & \\
\text{s.t.} & 2x_3 & + & x_4 & \leq & 2 \\
 & & & 2x_4 & \leq & 3 \\
 & x_3 & \geq & 0, x_4 & \geq & 0
\end{array}$$

$$\begin{array}{rrrrcl}
\max & 2x_5 & + & x_6 & & \\
\text{s.t.} & & & x_6 & \leq & 1 \\
 & x_5 & + & x_6 & \leq & 1 \\
 & x_5 & \geq & 0, x_6 & \geq & 0
\end{array}$$

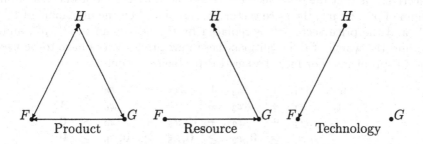

Figure 2.2 The transfer graphs.

Solving these three problems indicates that the optimal production plan for Inez is to produce nothing at Freetown, to produce 0.25 unit of product p_3 and 1.5 units of product p_4 at Garret, and to produce 1 unit of product p_5 at Hyattsville. Then her profit will be 0+1.75+2=3.75. The possibility of transferring products, resources, and technology between the three sites is added to the model. With three directed graphs we can describe which transfers are possible between which sites. We will consider these transfers one after another starting with the technology transfer. Successive transfers along more than one arc are not allowed. In figure 2.2 we see that technology transfer is only possible from Hyattsville to Freetown. By transferring technology from Hyattsville to Freetown it becomes possible to produce products p_5 and p_6 at Freetown using production matrix A^H. So we obtain the new production matrix at Freetown by adding the rows of A^H to A^F. We can again compute the worth of each coalition by considering three separate LP-problems. The LP-problems for Garret and Hyattsville do not change. For Freetown the new production matrix has to be used. For the coalition consisting only of Inez this yields the following LP-problem:

$$
\begin{array}{rlllllll}
\max & x_1 & + & 2x_2 & + & 2x_5 & + & x_6 \\
\text{s.t.} & x_1 & + & 3x_2 & + & & & x_6 & \leq & 4 \\
& 2x_1 & + & x_2 & + & x_5 & + & x_6 & \leq & 0 \\
& x_1 & \geq & 0, x_2 & \geq & 0, x_5 & \geq & 0, x_6 & \geq & 0.
\end{array}
$$

The optimal solution for Inez is the same as in the case without technology transfer.

Adding product transfers to the model implies that a product made at one location can be sold at another location. In figure 2.2 we see that product transfers are possible from Freetown to Hyattsville, from Garret to Freetown, from Hyattsville to Freetown, and from Hyattsville to Garret. What this means, effectively, is that the price vector p^F is replaced by the (coordinate-wise) maximum of p^F and p^H, the price vector p^G is replaced by the maximum of p^G and p^F, and the price vector p^H is replaced by the maximum of p^H, p^F, and p^G. To find the worth of the coalitions these new price vectors have to be used in the LP-problems. For Inez the new LP-problems become:

$$
\begin{array}{rlllll}
\max & 3x_1 + & 2x_2 + & 2x_5 + & x_6 & \\
\text{s.t.} & x_1 + & 3x_2 + & & x_6 & \leq 4 \\
& 2x_1 + & x_2 + & x_5 + & x_6 & \leq 0 \\
& x_1 \geq & 0, x_2 \geq & 0, x_5 \geq & 0, x_6 & \geq 0
\end{array}
$$

$$
\begin{array}{rll}
\max & 3x_3 + & x_4 \\
\text{s.t.} & 2x_3 + & x_4 \leq 2 \\
& & 2x_4 \leq 3 \\
& x_3 \geq & 0, x_4 \geq 0
\end{array}
\qquad
\begin{array}{rll}
\max & 2x_5 + & x_6 \\
\text{s.t.} & & x_6 \leq 1 \\
& x_5 + & x_6 \leq 1 \\
& x_5 \geq & 0, x_6 \geq 0.
\end{array}
$$

The optimal solution for Inez in this case is to produce nothing at Freetown, to produce 1 unit of product p_3 at Garret, and to produce 1 unit of product p_5 at Hyattsville. This will give her a profit of $0+3+2=5$.

Finally, we introduce the possibility of resource transfers in the model. In figure 2.2 we see that resource transfers are possible from Freetown to Garret and from Garret to Hyattsville. (Remember that successive transfers along two or more arcs are not allowed). This situation cannot be analyzed with three separate LP-problems. We have to consider one large LP-problem with extra variables to denote the transfers of resources. Let r_k^{FG} (r_k^{GH}) denote the amount of resource k that is transferred from Freetown to Garret (Garret to Hyattsville). For Inez this LP-problem is given below.

$$
\begin{aligned}
\max \quad & 3x_1 + 2x_2 + 2x_5 + x_6 + 3y_3 + y_4 + 2z_5 + z_6 \\
\text{s.t.} \quad & x_1 + 3x_2 + x_6 \leq 4 - r_1^{FG} \\
& 2x_1 + x_2 + x_5 + x_6 \leq 0 - r_2^{FG} \\
& 2y_3 + y_4 \leq 2 + r_1^{FG} - r_1^{GH} \\
& 2y_4 \leq 3 + r_2^{FG} - r_2^{GH} \\
& z_6 \leq 1 + r_1^{GH} \\
& z_5 + z_6 \leq 1 + r_2^{GH} \\
& 0 \leq r_1^{FG} \leq 4 \\
& 0 \leq r_2^{FG} \leq 0 \\
& 0 \leq r_1^{GH} \leq 2 \\
& 0 \leq r_2^{GH} \leq 3 \\
& x_1, x_2, x_5, x_6, y_3, y_4, z_5, z_6 \geq 0
\end{aligned}
$$

Here x_i is the amount of product p_i that Inez makes at Freetown, y_i is the amount of product p_i that Inez makes at Garret, and z_i is the amount of product p_i that Inez makes at Hyattsville.

Solving this LP-problem gives Inez the following optimal production plan: produce nothing at Freetown, produce 3 units of product p_3 at Garret, and produce 4 units of product 5 at Hyattsville. To do this she has to transfer 4 units of resource 1 from Freetown to Garret, and three units of resource 2 from Garret to Hyattsville. This will give her a profit of 0+9+8=17.

Similarly the worths of all other coalitions can be computed.

One can show in a straightforward way that linear production games with transfers are linear programming games and hence, they are totally balanced. Further generalizations of this model can be obtained by introducing *license fees* that have to be paid when a technology is transferred from one site to another; by considering transfer costs for products as well as resources; and by considering losses that occur when products or resources are transferred. The games that arise remain linear programming games and therefore, totally balanced.

Owen [91] showed for linear production games that not all core elements can be obtained from the solutions of the dual problem. However, if the players are replicated, then the core will converge to the set of payoffs generated by the optimal solutions of the dual problem. Owen shows that if the dual problem has a unique solution, the convergence is finite.

Samet and Zemel [113] give a necessary and sufficient condition for finite convergence of the core of a linear programming game to the set of payoff vectors corresponding to optimal solutions of the dual problem.

Dubey and Shapley [41] consider totally balanced games arising from, not necessarily linear, programming problems with constraints controlled by the players.

2.3 LINEAR PROGRAMMING GAMES WITH COMMITTEE CONTROL

The concept of committee control in linear programming games is best illustrated with two examples. Consider a flow game with graph G as described in the previous section with the exception of the ownership function. Instead of having one owner, the arcs are controlled by committees consisting of subsets of players. This committee control is modelled with simple games. To every arc $l \in L$ we assign a simple game w_l. A coalition S is said to control l if and only if $w_l(S) = 1$. The graph G_S is obtained from G by keeping all the vertices and removing all arcs that are not controlled by S. We define the worth $v(S)$ of coalition S to be the value of a maximum flow in G_S. Contrary to ordinary

source  sink

Figure 2.3 A flow game with committee control with an empty core.

flow games, flow games with committee control need not have an empty core as the following simple example shows.

Example Let $N = \{1, 2\}$, $P=\{$source, sink$\}$, $L = \{l_1\}$, with $B(l_1) =$source, $E(l_1) =$sink, $c(l_1) = 10$, and $w_l(S) = 10$ for every $S \in 2^N \setminus \{\emptyset\}$. See figure 2.3. For this game we have $v(S) = 10$ for every S, so $C(v) = \emptyset$. Of course, flow games with an ownership function can also be considered flow games with committee control, where the owner of an arc is the dictator in the simple game that describes the control of that arc.

Another example of linear programming games with committee control is given by linear production games where the resources are controlled by committees of players. Suppose that the resources in a linear production situation are available in portions. For $k \in \{1, \ldots, m\}$ there are g_k portions of resource k. Such a portion is denoted by B_k^q. The total amount of resource k that is available is $\sum_{q=1}^{g_k} B_k^q$. Each portion is controlled by committees of players. For every B_k^q a simple game w_k^q describes the control. Coalition S can use B_k^q only if $w_k^q(S) = 1$. Let $B_k^q(S) := B_k^q w_k^q(S)$. The total amount of the k-th resource available to coalition S is $B_k(S) := \sum_{q=1}^{g_k} B_k^q(S)$. The resource vector $B(S)$ of coalition S is the vector $(B_1(S), B_2(S), \ldots, B_m(S))$. Then $v(S)$ is the value of problem 2.5 with $b(S)$ replaced by $B(S)$. The following example shows that a linear production game with committee control can have an empty core.

Example Consider the production situation with two resources and two products p_1, p_2. The production matrix A and price vector p are given below.

$$A = \begin{pmatrix} 1 & 1 \\ 0 & 1 \end{pmatrix} p = \begin{pmatrix} 6 \\ 5 \end{pmatrix}$$

There are available 100 units of resource 1 and 110 units of resource 2. Larry alone controls 50 units of resource 1, Mary alone controls 30 units of resource 1, and Mary and Nick together control 20 units of resource 1. Nick controls 20 units of resource 2, and any majority of the three players has control over 90 units of resource 2. Using the notation introduced above and Larry=1,

S	B(S)	v(S)	S	B(S)	v(S)	S	B(S)	v(S)
{1}	(50,0)	0	{1,2}	(80,90)	530	{1,2,3}	(100,110)	650
{2}	(30,0)	0	{1,3}	(50,110)	600	∅	(0,0)	0
{3}	(0,20)	100	{2,3}	(50,110)	600			

Table 2.3 The resource vectors and values of the coalitions in a linear production game with committee control.

Mary=2, Nick=3, we obtain

$$B_1^1 = 50 \quad MW(w_1^1) = \{\{1\}\}$$
$$B_1^2 = 30 \quad MW(w_1^2) = \{\{2\}\}$$
$$B_1^3 = 20 \quad MW(w_1^3) = \{\{2,3\}\}$$
$$B_2^1 = 20 \quad MW(w_2^1) = \{\{3\}\}$$
$$B_2^2 = 90 \quad MW(w_2^2) = \{\{1,2\},\{1,3\},\{2,3\}\}$$

Here $MW(w)$ denotes the collection of minimal winning coalitions of the simple game w. Since the simple games in this example are monotonic, they are completely described by the minimal winning coalitions. In table 2.3 the resource vectors and values of the coalitions are given. Since

$$v(1,2) + v(1,3) + v(2,3) = 1730 > 1300 = 2v(N)$$

it follows that v is not balanced.

In general it is possible to consider linear programming games with committee control.

Definition 2.3.1 *A game v is a linear programming game with committee control if $v(S) = v_p(A, H, b(S), d(S), c)$ where A is an $m \times p$-matrix, H is an $m \times r$-matrix, $c \in R^m$, $b_k(S) = \sum_{q=1}^{g_k} b_k^q w_k^q(S)$ for all $k \in \{1, \ldots, p\}$, and $d_j(S) = \sum_{q=1}^{h_j} d_j^q u_j^q(S)$. Here $b_k^1, b_k^2, \ldots, b_k^{g_k}$ and $d_j^1, d_j^2, \ldots, d_j^{h_j}$ are real non-negative numbers, and $w_k^1, \ldots, w_k^{g_k}$ and $u_j^1, \ldots, u_j^{h_j}$ are simple games.*

Theorem 2.3.2 *Let v be a linear programming game with committee control such that all the simple games that describe the controls have veto players. Then $C(v)$ is not empty.*

Proof. Let $(\hat{y}, \hat{z})$ be an optimal solution for the dual problem of the problem that determines $v(N)$. Since all the simple games involved have veto players each one has a non-empty core. Let $\eta_k^q \in C(w_k^q)$ and $\zeta_j^q \in C(u_j^q)$. Define $x \in R^n$ by

$$x_i := \sum_{k=1}^{p} \hat{y}_k \sum_{q=1}^{g_k} b_k^q (\eta_k^q)_i + \sum_{j=1}^{r} \hat{z}_j \sum_{q=1}^{h_j} d_j^q (\zeta_j^q)_i \, .$$

Then

$$\begin{aligned}
x(N) &= \sum_{k=1}^{p} \hat{y}_k \sum_{q=1}^{g_k} b_k^q + \sum_{j=1}^{r} \hat{z}_j \sum_{q=1}^{h_j} d_j^q \\
&= \sum_{k=1}^{p} \hat{y}_k b_k^q(N) + \sum_{j=1}^{r} \hat{z}_j d_j^q(N) \\
&= \hat{y}b(N) + \hat{z}d(N) = v(N).
\end{aligned}$$

Further,

$$\begin{aligned}
x(S) &= \sum_{k=1}^{p} \hat{y}_k \sum_{q=1}^{g_k} b_k^q \eta_k^q(S) + \sum_{j=1}^{r} \hat{z}_j \sum_{q=1}^{h_j} d_j^q \zeta_k^q(S) \\
&\geq \sum_{k=1}^{p} \hat{y}_k \sum_{q=1}^{g_k} b_k^q w_k^q(S) + \sum_{j=1}^{r} \hat{z}_j \sum_{q=1}^{h_j} d_j^q u_j^q(S) \\
&= \sum_{k=1}^{p} \hat{y}_k b_k(S) + \sum_{j=1}^{r} \hat{z}_j d_j(S) \\
&= \hat{y}b(S) + \hat{z}d(S) \geq v(S).
\end{aligned}$$

Here the last inequality follows from the fact that $(\hat{y}, \hat{z})$ is a solution of the dual problem that determines $v(S)$ as well. Thus, it follows that $x \in C(v)$. $\qquad\square$

For flow games with committee control we obtain a core element by distributing the capacities of each arc of a minimum cut among the veto players of the simple game that describes the control of that arc. A *cut* in a directed graph G with a source and a sink is defined as follows. Let A be a subset of P such that the sink is an element of A, and the source is an element of $P \setminus A$. By $(A, P \setminus A)$ we denote the subset of L consisting of all arcs with a starting point in A and an endpoint in $P \setminus A$. Formally,

$$(A, P \setminus A) := \{l \in L | l \in B(p) \text{ for a } p \in A \text{ and } l \in E(q) \text{ for a } q \in P \setminus A\}.$$

Such a subset of L is called a cut of G. Note that a cut of G can be made into a cut of G_S by removing all arcs which are not controlled by S from it. The

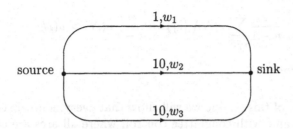

Figure 2.4 A balanced flow game with committee control.

capacity of a cut is the sum of the capacities of its arcs. A cut with minimum capacity is called a *minimum cut*. The well-known max flow-min cut theorem of Ford and Fulkerson states that the value of a maximum flow is equal to the value of a minimum cut.

It follows that for flow games with committee control it is not necessary for all the simple games involved to have veto players to obtain the balancedness of the game. It is sufficient that there exists a minimum cut with all its arcs controlled by simple games with veto players. However, the following example shows that even this is not a necessary condition.

Example Let $N = \{1, 2, 3, 4\}$. The graph G is given in figure 2.4. The numbers denote the capacities of the arcs. The winning coalitions of w_1 are $\{1,3\}, \{2,4\}$, and N. The winning coalitions of w_2 are $\{1,2\}$ and N. The winning coalitions of w_3 are $\{3,4\}$ and N. Note that there is no minimum cut with all its arcs having control games with veto players. The core of the game, however is not empty. An element of the core is, for example, $(6,5,5,5)$.

Theorem 2.3.3 *A linear programming game v, with $v(N) > 0$, with committee control for which all the simple games that describe the controls are monotonic and non-balanced has an empty core.*

Proof. Let v, with $v(N) > 0$, be a linear programming game with committee control described by simple games w_k^q and u_k^q. Because w_k^q (u_k^q) is monotonic without veto players it follows that $w_k^q(N \setminus \{i\}) = w_k^q(N)$ ($u_k^q(N \setminus \{i\}) = u_k^q(N)$) for all $i \in N$. So $b(N \setminus \{i\}) = b(N)$, $d(N \setminus \{i\}) = d(N)$ and therefore, $v(N \setminus \{i\}) = v(N)$ for all $i \in N$. Consider the balanced collection $\{N \setminus \{i\}\}_{i \in N}$

with all weights equal to 1/(n-1). Then

$$\frac{1}{n-1}\sum_{i\in N} v(N\setminus\{i\}) = \frac{n}{n-1}v(N) > v(N).$$

So v is not balanced.   □

In the remainder of this section we will show that every non-negative balanced game is a flow game with committee control, where all arcs are controlled by simple games with veto players. In the following we will call such games *veto rich flow games*. This result was first proven in [19].
Let $v_1, v_2 \in G^N$. Then $v_1 \wedge v_2$ is the game defined by
$v_1 \wedge v_2(S) = \min\{v_1(S), v_2(S)\}$ for all $S \in 2^N$. Let $\alpha, \beta \in R$. Then the game $\alpha v_1 + \beta v_2$ is defined by $\alpha v_1 + \beta v_2(S) = \alpha v_1(S) + \beta v_2(S)$.

Lemma 2.3.4 *Let v_1 and v_2 be two veto rich flow games. Then the games $< N, v_1 \wedge v_2 >$ and $< N, \alpha v_1 + \beta v_2 >$ with $\alpha, \beta \geq 0$, are also veto rich flow games.*

Proof. Let G_i for $i = 1, 2$ be the graph from which v_i is obtained. Then $v_1 \wedge v_2$ is obtained from the graph which results when we combine G_1 and G_2 in series, hereby identifying the sink of G_1 with the source of G_2.
Multiply all capacities in G_1 by α and all capacities in G_2 by β. Then $\alpha v_1 + \beta v_2$ is obtained from the graph which results by combining G_1 and G_2 with their new capacities in parallel, hereby identifying the sink of G_1 with the sink of G_2 and the source of G_1 by the source of G_2.   □

Theorem 2.3.5 *Every non-negative balanced game is a veto rich flow game.*

Proof. Let v be a non-negative balanced game. Then there exist balanced simple games $w_1, w_2, \ldots, w_k$, and $\alpha_1, \alpha_2, \ldots, \alpha_k \geq 0$, such that $v = \alpha_1 w_1 + \alpha_2 w_2 + \ldots + \alpha_k w_k$. Trivially, each w_j is a veto rich flow game with one arc with capacity 1 and with committee control described by w_j. With lemma 2.3.4 it follows that v is a veto rich flow game.   □

Since flow games with committee control are linear programming games with committee control, theorem 2.3.5 shows that every balanced game is a linear programming game with committee control. This can also be shown directly by observing that $v(S) = v_p(A, I, b(S), d(S), c)$ where A is the 1×1-zero matrix,

I is the 1×1-identity matrix, $b(S) = 0$, $d(S) = \sum_{j=1}^{p} \alpha_j w_j(S)$ and $c = 1$. Here the w_j's and α_j's are as given in the proof of theorem 2.3.5.

2.4 NON-BALANCED LINEAR PROGRAMMING GAMES

In this section we will discuss some ways of allocating revenues in non-balanced linear programming games, that appear in [22].

Let us consider the game of table 2.3. Since the game is not balanced there is no stable allocation. Any proposed payoff vector will induce a coalition to leave. Still the total profit is maximal if everybody cooperates. The LP-problem that determines $v(N)$ is

$$
\begin{array}{rrcll}
\max & 6x_1 + & 5x_2 & & \\
\text{s.t.} & x_1 & & \leq & 100 \\
& x_1 + & x_2 & \leq & 110 \\
& x_1 \geq & 0, x_2 & \geq & 0.
\end{array}
$$

The dual problem is

$$
\begin{array}{rrcll}
\min & 100y_1 + & 110y_2 & & \\
\text{s.t.} & y_1 + & y_2 & \geq & 6 \\
& & y_2 & \geq & 5 \\
& y_1 \geq & 0, y_2 & \geq & 0.
\end{array}
\tag{2.6}
$$

The solution of problem 2.6 is $(1,5)$. The shadow price of resource 1 is 1, that of resource 2 is 5. Resource 1 does not pose a problem. Larry can receive a payoff of 50 for the 50 units of resource 1 that he owns, and Mary can receive a payoff of 30 for the 30 units of resource 2 that she owns. The payoff of 20 for the remaining 20 units can be divided equally between Mary and Nick since they are both veto players in the game that controls these units. The 20 units of resource 2 owned by Nick do not form a problem also. Nick receives a payoff of 100 for these. The problem is created by the 90 units of resource 2 that are controlled by any majority of the three players. So it is for these that a solution should be created. One way of doing this is by levying a tax on the formation of any coalition other than the grand coalition. This can be done by a regulating body that wants to create incentives for everybody to cooperate in the case that this course of action is the most profitable. This body will say to Larry and Mary: "Do you want to form a coalition without Nick? Okay, but then you will have to pay a tax which depends on the added value that you create by forming this coalition." The *added value* of coalition $\{1,2\}$ in the game w_2^2, that controls the 90 units of resource 2, is $v(1,2) - v(1) - v(2) = 1 - 0 - 0 = 1$, the added value of coalition $\{1,3\}$ in w_2^2 is $v(1,3) - v(1) - v(3) = 1 - 0 - 0 = 1$. In the same way the added value of coalition $\{2,3\}$ is computed to be 1. The tax imposed will be proportional to these added values. Its aim is to create a balanced game by lowering the values of all coalitions except the grand coalition and the one-person coalitions. The tax rate will be the lowest one that achieves this. In this example that is $1/3$. Instead of w_2^2 we get the game $w^{1/3}$ with $w^{1/3}(i) = 0$ for $i = 1,2,3$, $w^{1/3}(i,j) = 2/3$ for $i,j \in \{1,2,3\}$, and $w^{1/3}(N) = 1$. Effectively this means that two-person coalitions control only $2/3$ of the 90 units of resource 2 instead of 100%. This game has a unique core element $(1/3,1/3,1/3)$. The payoff of 450 created by the 90 units of resource 2 can be divided with this core element as division key, i.e., every player receives 150. So the total payoff to Larry will be 200, to Mary 190, and to Nick 260. Let us formalize this way of arriving at an allocation of the profit in a non-balanced linear programming game.

Let v be a game such that $v(N) \geq \sum_{i \in N} v(i)$. Then each player can be guaranteed at least an individual rational payoff. If this is not the case then the players will be better off by not forming the grand coalition. The added value created by the formation of coalition S is $v(S) - \sum_{i \in S} v(i)$. A tax, proportional to this is imposed on every coalition $s \neq N$. So this tax equals

$$\varepsilon(v(S) - \sum_{i \in S} v(i)) \text{ for an } 0 \leq \varepsilon \leq 1.$$

Thus, a new game, the *multiplicative ε-tax game* is constructed.

Definition 2.4.1 *The multiplicative ε-tax game v^ε of a game v is given by*

$$v^\varepsilon(S) := \begin{cases} v(S) & \text{if } S = N, \emptyset \\ v(S) - \varepsilon(v(S) - \sum_{i \in S} v(i)) & \text{otherwise.} \end{cases}$$

Now we consider the smallest ε for which $C(v^\varepsilon) \neq \emptyset$. If $C(v) \neq \emptyset$ then this ε is equal to 0. For $\varepsilon = 1$ the multiplicative ε-tax game is given by $v^1(S) = \sum_{i \in S} v(i)$ for $S \neq N, \emptyset$. So $C(v^1) \neq \emptyset$. For every game v with $\sum_{i \in N} v(i) \leq v(N)$ define

$$\varepsilon(v) := \min\{\varepsilon | 0 \leq \varepsilon \leq 1, \ C(v^\varepsilon) \neq \emptyset\}.$$

Then $\varepsilon(v)$ is well defined since the balanced games form a convex polyhedral cone and $C(v^1) \neq \emptyset$.

Definition 2.4.2 *The least tax core $LTC(v)$ of the game v is the core of the game $v^{\varepsilon(v)}$.*

Let w be a simple game with an empty core and with $w(N) \geq \sum_{i \in N} w(i)$. We can distinguish two cases.

1. There is an $i^* \in N$ such that $w(i^*) = 1$.

2. For all $i \in N$ we have $w(i) = 0$.

In the first case there is an $S \subset N \setminus \{i^*\}$ with $w(S) = 1$ since i^* is not a veto player. For $\varepsilon < 1$ we have $w^\varepsilon(S) = 1 - \varepsilon > 0$ and $w^\varepsilon(i^*) = 1$ and it is evident that $C(w^\varepsilon) = \emptyset$. Hence $\varepsilon(w) = 1$ and

$$w^{\varepsilon(w)}(S) = \begin{cases} 1 & \text{if } i^* \in S \\ 0 & \text{if } i^* \notin S. \end{cases}$$

In the second case we have

$$w^{\varepsilon(w)} = \begin{cases} 0 & \text{if } w(S) = 0 \\ 1 - \varepsilon(w) & \text{if } w(S) = 1, \ S \neq N. \end{cases}$$

In the first case $w^{\varepsilon(w)}$ is still a simple game and i^* has become a unique veto player. In the second case $w^{\varepsilon(w)}$ is no longer a simple game. We can interpret committee control as described by the game $w^{\varepsilon(w)}$ in the following way. Let w be the simple game which describes the control of a constraint b. Then $w^{\varepsilon(w)}$ can be seen as describing the situation where only part of b is controlled by a coalition S with $w(S) = 1$, namely, the part given by $(1 - \varepsilon(w))b$.

Let v be a linear programming game with committee control given by simple games w_k^q and u_k^q. Then the *least tax linear programming game* corresponding to v is the linear programming game with controls described by the least tax games constructed from the simple games w_k^q and u_k^q.

Theorem 2.4.3 *Let v be a linear programming game with committee control. Then the corresponding least tax linear programming game is balanced.*

Proof. Similar to the proof of theorem 2.3.2 by taking elements of the least tax core of the simple games which describe the controls. $\square$

Let us look at a slight modification of the game of table 2.3. The 90 units of resource 1 are now not only controlled by any majority of the three players but also by Larry alone. The least tax core will assign all the payoff of 450 created by these 90 units to Larry, ignoring the fact that Mary and Nick together can

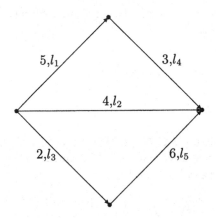

Figure 2.5   The graph of a flow game with claims.

also control them. If, instead, we compute the Shapley-Shubik power index for this new control game we obtain a division key of $(2/3,1/6,1/6)$, leading to the allocation $(300,75,75)$. The (normalized) Banzhaf index yields the division key $(3/5,1/5,1/5)$, leading to the allocation $(270,90,90)$. This illustrates an alternative way of arriving at a payoff vector when committee control is involved. A normalized power index ψ can be used to compute the power of each player in the controlling committees. If $\psi_i(w_k^q)$ is a measure of the power of player i in the control game w_k^q which describes the control of b_k^q, then this can be regarded as if player i controls $\psi_i(w_k^q)b_k^q$. Let $(\hat{y},\hat{z})$ be an optimal solution for the dual of the problem which determines $v(N)$. Then $v(N)$ can be divided by giving to player i the amount $\sum_{k=1}^{p} \hat{y}_k \sum_{q=1}^{g_k} \psi_i(w_k^q) + \sum_{j=1}^{r} \hat{z}_j \sum_{q=1}^{h_j} \psi_i(u_k^q)$.

2.5   LINEAR PROGRAMMING GAMES WITH CLAIMS

Olga, Peter, Quirina, and Roald all claim that they have the right to use a certain amount of the capacities of the arcs in the graph of figure 2.5. Olga, for example, claims 2 units of arc l_1, 1 unit of arc l_2, 1 unit of arc l_3, 2 units of arc l_4, and 2 units of arc l_5. The claims of all four of them are given in table 2.4. Since for each arc the sum of the claims of the players is more than the capacity of the arc it is impossible to assign to every player her/his claim. For each arc l_i we can construct a *claim game* v^{l_i} to describe the situation. The coalitions in these claim games adapt a somewhat pessimistic view of the

	l_1	l_2	l_3	l_4	l_5
Olga	2	1	1	2	2
Peter	3	3	1	1	3
Quirina	1	0	1	2	3
Roald	0	2	1	1	3

Table 2.4 The claims of the players on the arcs of figure 2.5.

S	$v^{l_1}(S)$	$v^{l_2}(S)$	$v^{l_3}(S)$	$v^{l_4}(S)$	$v^{l_5}(S)$	v(S)
{1}	1	0	0	0	0	0
{2}	2	1	0	0	0	0
{3}	0	0	0	0	0	0
{4}	0	0	0	0	0	0
{1,2}	4	2	0	0	0	2
{1,3}	2	0	0	1	0	1
{1,4}	1	1	0	0	0	1
{2,3}	3	1	0	0	1	1
{2,4}	2	3	0	0	1	3
{3,4}	0	0	0	0	1	0
{1,2,3}	5	2	1	2	3	5
{1,2,4}	4	4	1	1	3	6
{1,3,4}	2	1	1	2	3	4
{2,3,4}	3	3	1	1	4	5
{1,2,3,4}	5	4	2	3	6	9

Table 2.5 Claim games and a flow game with claims.

situation. Coalition S assumes that it can receive, without problems, that part of the capacity of an arc that is not claimed by $N \setminus S$. These claim games determine how much of the capacity of each arc a coalition can use. With this, it is then possible to compute the characteristic function v of the flow game. The claim games v^{l_i} and the flow game v are given in table 2.5. Here Olga=1, Peter=2, Quirina=3 and Roald=4. A minimum cut of the graph of figure 2.5 is $\{l_2, l_3, l_4\}$. A core element of v^{l_2} is $(1,2,0,1)$, a core element of v^{l_3} is $(1,0,1,0)$, and a core element of v^{l_4} is $(1,1,1,0)$. Combining these yields the core element $(3,3,2,1)$ of v.

Sarah, Tom, and Ursula can make two products, product p_1 and product p_2, using three types of resources r_1, r_2, r_3 which are available in the amounts 50,40,60, respectively. The production matrix A and price vector p are given

S	$v^1(S)$	$v^2(S)$	$v^3(S)$	$v(S)$
$\{1\}$	0	0	0	0
$\{2\}$	0	0	0	0
$\{3\}$	0	0	0	0
$\{1,2\}$	10	10	20	14
$\{1,3\}$	16	12	24	20
$\{2,3\}$	14	18	28	22
$\{1,2,3\}$	50	40	60	64

Table 2.6   A linear production game with claims.

below.

$$A = \begin{pmatrix} 3 & 1 & 2 \\ 1 & 2 & 2 \end{pmatrix} \quad p = \begin{pmatrix} 3 \\ 2 \end{pmatrix}$$

Each one of the three claims as much of each resource as she/he can use. Sarah knows that she can sell not more than 10 units of p_1 and not more than 6 units of p_2. Tom knows that he can sell not more than 8 units of p_1 and not more than 10 units of p_2. Ursula knows that she can sell not more than 10 units of each of the two products. To determine how much of each resource she will claim, Sarah solves the following linear programming problem:

$$
\begin{array}{rrcrcl}
\max & 3x_1 & + & 2x_2 & & \\
\text{s.t.} & 3x_1 & + & x_2 & \leq & 50 \\
& x_1 & + & 2x_2 & \leq & 40 \\
& 2x_1 & + & 2x_2 & \leq & 60 \\
& 0 \leq & x_1 & & \leq & 10 \\
& 0 \leq & x_2 & & \leq & 6
\end{array}
$$

The optimal solution is to produce 10 units of p_1 and 6 units of p_2. To do this she needs 36 units of r_1, 22 of r_2 and 32 of r_3. This is what she claims. Similarly Tom finds out that the best thing for him to do is to produce 8 units of p_1 and 10 units of p_2. To do this he claims 34 units of r_1, 28 units of r_2, and 36 units of r_3. Ursula claims 40 units of r_1, 30 units of r_2 and 40 units of r_3 to produce 10 units of p_1 and 10 units of p_2. Similarly as in the previous example these claims lead to claim games v^1, v^2, v^3, for each resource. In table 2.6 the claim games v^1, v^2, v^3, and the linear production game v are given. Here Sarah=1, Tom=2, and Ursula=3. The way to determine $v(1, 2)$, for example, is to solve the linear programming problem

$$
\begin{array}{rlrll}
\max & 3x_1 & + & 2x_2 & \\
\text{s.t.} & 3x_1 & + & x_2 & \leq & 10 \\
& x_1 & + & 2x_2 & \leq & 10 \\
& 2x_1 & + & 2x_2 & \leq & 20 \\
& 0 & \leq & x_1 & \leq & 18 \\
& 0 & \leq & x_2 & \leq & 16
\end{array}
$$

The worth of the grand coalition is given by the linear programming problem

$$
\begin{array}{rlrll}
\max & 3x_1 & + & 2x_2 & \\
\text{s.t.} & 3x_1 & + & x_2 & \leq & 50 \\
& x_1 & + & 2x_2 & \leq & 40 \\
& 2x_1 & + & 2x_2 & \leq & 60 \\
& x_1 & \geq & 0, x_2 & \geq & 0
\end{array}
\tag{2.7}
$$

The dual problem of problem 2.7 is

$$
\begin{array}{rlrlrll}
\min & 50y_1 & + & 40y_2 & + & 60y_3 & \\
\text{s.t.} & 3y_1 & + & y_2 & + & 2y_3 & \geq & 3 \\
& y_1 & + & 2y_2 & + & 2y_3 & \geq & 2 \\
& y_1 & \geq & 0, y_2 & \geq & 0, y_3 & \geq & 0
\end{array}
\tag{2.8}
$$

The optimal solution of problem 2.8 is (0.8,0.6,0). This can be combined with core elements of the claim games to yield a core element of the linear production game v. For example, the core elements (15,15,20), (20,10,10), (20,20,20) of v^1, v^2, v^3, respectively, lead to the core element (24,18,22) of v.

The flow game and the linear production game discussed above are examples of linear programming games with claims. In an LP-game with claims each player claims a certain part of the right hand sides of the constraints. The amount that player i claims of each constraint depends on the optimal solution of the following LP-problem.

$$
\begin{array}{rl}
\max & c \cdot x \\
\text{s.t.} & xA \leq B \\
& xH = D \\
& 0 \leq x \leq k(i)
\end{array}
\tag{2.9}
$$

Here A is an $m \times p$-matrix, H is an $m \times r$-matrix, $B \in R^p$, $D \in R^r$, $c \in R^m$, and $k(i) \in R^m$ for each $i \in N$. Let x^i be the (unique)[2] optimal solution of problem 2.9. Then player i claims $b(i) = x^i A$ of B and $d(i) = x^i H$ of D. A

---

[2]If there are more optimal solutions some way of choosing one of them to determine the claims should be specified. The way this is done is not important for the remainder of the analysis.

coalition S claims $b(S) := \sum_{i \in S} b(i)$ of B and $d(S) := \sum_{i \in S} d(i)$ of D. In determining $v(S)$ in an LP-game with claims we assume that S has access to $(B - b(N \setminus S))_+ := \max\{B - b(N \setminus S), 0\}$ units of B and $(D - d(N \setminus S))_+$ units of D, and that the restriction on x is $k(S) := \sum_{i \in S} k(i)$.

Definition 2.5.1 *In a linear programming game with claims, $v(S)$ is equal to the value of the linear programming problem*

$$
\begin{array}{rl}
\max & c \cdot x \\
& xA \leq (B - b(N \setminus S))_+ \\
& xH \leq (D - d(N \setminus S))_+ \\
& 0 \leq x \leq k(S).
\end{array}
\tag{2.10}
$$

Definition 2.5.2 *For each B_j (D_j) let v^j (u^j) be the claim game given by*

$$
\begin{array}{rll}
v^j(S) & = & (B_j - b_j(N \setminus S))_+ \text{ for each } S \in 2^N \setminus \emptyset \\
(u^j(S) & = & (D_j - d_j(N \setminus S))_+ \text{ for each } S \in 2^N \setminus \emptyset).
\end{array}
$$

Theorem 2.5.3 *The claim games defined in 2.5.2 are convex.*

Proof. Let v^j be a claim game as defined in 2.5.2. Let $T \subset S \subset N \setminus \{i\}$. From theorem 1.4.2 it follows that it is sufficient to show that

$$v^j(S \cup \{i\}) + v^j(T) \geq v^j(T \cup \{i\}) + v^j(S).$$

Let $F := B_j - \sum_{i \in} b_j(i)$. Then

$$
\begin{array}{rll}
v^j(S \cup \{i\}) + v^j(T) & = & \max\{F + b_j(S) + b_j(i), 0\} + \max\{F + b_j(T), 0\} \\
& = & \max\{2F + b_j(S) + b_j(T) + b_j(i), F + b_j(S) + b_j(i), \\
& & F + b_j(T), 0\}
\end{array}
$$

and

$$
\begin{array}{rll}
v^j(T \cup \{i\}) + v^j(S) & = & \max\{F + b_j(T) + b_j(i), 0\} + \max\{F + b_j(S), 0\} \\
& = & \max\{2F + b_j(S) + b_j(T) + b_j(i), F + b_j(T) + b_j(i), \\
& & F + b_j(S), 0\}.
\end{array}
$$

Since $b_j(S) \geq b_j(T)$ it follows that $F + b_j(T) + b_j(i) \leq F + b_j(S) + b_j(i)$. Further, $F + b_j(S) \leq F + b_j(S) + b_j(i)$. Therefore, $v^j(S \cup \{i\}) + v^j(T) \geq v^j(T \cup \{i\}) + v^j(S)$. $\square$

Now we can proof the following theorem.

Theorem 2.5.4 *Linear programming games with claims have a non-empty core.*

Proof. Let v be a linear programming game with claims. Then all the claim games are convex and therefore balanced. Consider the dual problem of 2.10 for $S = N$. That is

$$\begin{aligned}
\min \quad & B \cdot y + D \cdot z + k(N) \cdot w \\
\text{s.t.} \quad & Ay + Hz + w \geq c \\
& y, w \geq 0.
\end{aligned} \qquad (2.11)$$

Let $(\hat{y}, \hat{z}, \hat{w})$ be an optimal solution for 2.11. Let $q^j \in C(v^j)$ and $s^j \in C(u^j)$. Define $x \in R^n$ by

$$x_i := \sum_{j=1}^{p} q_i^j \hat{y}_j + \sum_{j=1}^{r} s_i^j \hat{z}_j + \sum_{j=1}^{m} k_j(i) \hat{w}_j \ .$$

Then $x \in C(v)$. □

2.6 SIMPLE FLOW GAMES

In this section we consider a special class of flow games, the so called *simple flow games*. All the arcs in the underlying graph of a flow game have capacity 1 and each player owns exactly one arc. So the player set can be identified with the set of arcs. A *path* in the graph G of a simple flow game v is a minimal (with respect to inclusion) set of arcs that connect the source with the sink. A path can be identified with the coalition consisting of the players who own an arc in the path. Let SP denote the collection of coalitions which can be identified with a path in G. It is obvious that

$$C(v) = \{x \in R^n | x(N) = v(N), \ x_i \geq 0, \ x(S) \geq 1 \text{ for all } S \in SP\}.$$

For simple flow games it can be shown that every core element is generated by an optimal solution of the dual problem of the problem that determines $v(S)$. Let us illustrate this with an example before giving the formal proof.

Example In figure 2.6 the graph of a six-person simple flow game is given. It is easy to see that $v(N) = 3$. Any maximal flow f will have $f(1) = f(4) = f(5) = f(6) = 1$ and $f(2) + f(3) = 1$. There are two minimum cuts, namely

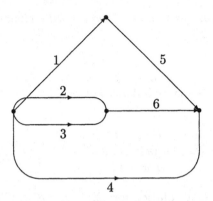

Figure 2.6 The graph of a simple flow game.

$\{4,5,6\}$ and $\{1,4,6\}$. The core elements generated by these two minimum cuts are $(0,0,0,1,1,1)$ and $(1,0,0,1,0,1)$, respectively. The core is the convex hull of these two vectors. The set $SP = \{\{1,5\}, \{2,6\}, \{3,6\}, \{4\}\}$. The core is equal to the set of optimal solutions of the LP-problem

$$
\begin{aligned}
\min \quad & u_1 + u_2 + u_3 + u_4 + u_5 + u_6 \\
\text{s.t.} \quad & u_1 + u_5 \geq 1 \\
& u_2 + u_6 \geq 1 \\
& u_3 + u_6 \geq 1 \\
& u_4 \geq 1 \\
& u \geq 0.
\end{aligned}
\tag{2.12}
$$

The LP-problem corresponding to $v(N)$ is

$$
\begin{aligned}
\max \quad & x_1 + x_2 + x_3 + x_4 \\
\text{s.t.} \quad & x_1 \leq 1 \\
& x_2 \leq 1 \\
& x_3 \leq 1 \\
& x_4 \leq 1 \\
& x_5 \leq 1 \\
& x_6 \leq 1 \\
& -x_1 + x_5 = 0 \\
& -x_2 - x_3 + x_6 = 0 \\
& x \geq 0.
\end{aligned}
\tag{2.13}
$$

The dual problem of 2.13 is

$$
\begin{array}{rrcl}
\min & y_1 + y_2 + y_3 + y_4 + y_5 + y_6 & & \\
\text{s.t.} & y_1 - z_1 & \geq & 1 \\
& y_2 - z_2 & \geq & 1 \\
& y_3 - z_2 & \geq & 1 \\
& y_4 & \geq & 1 \\
& y_5 + z_1 & \geq & 0 \\
& y_6 + z_2 & \geq & 0 \\
& y & \geq & 0.
\end{array}
\tag{2.14}
$$

Note the similarity between 2.14 and 2.12 which determines the core. If, in 2.14, we add the fifth constraint to the first one, and the sixth constraint to the second and third one, 2.14 becomes 2.12. The extreme points of the core are generated by the optimal solutions $(0,0,0,1,1,1,-1,-1)$ and $(1,0,0,1,0,1,0,-1)$ of 2.14. It follows that the whole core is generated by the set of optimal solutions of 2.14. Let $u \in R^6$ be an element of the core. Consider the LP-problem that results from 2.13 if we leave all the constraints the same and change the objective function from

$$(1,1,1,1,0,0) \cdot x$$

to

$$((1,1,1,1,0,0) - u) \cdot x.$$

Since u is a convex combination of the points $(0,0,0,1,1,1)$ and $(1,0,0,1,0,1)$ it follows that the new objective function is

$$(1 - u_1)x_1 + x_2 + x_3 - u_5 x_5 - x_6.$$

Since for any core element u we have $u_1 + u_5 = 1$ it follows that the value of the LP-problem with the new objective function is 0.

In the following we will show that the results we obtained in the example are valid in general. Consider the LP-problem that determines the core for simple flow games.

$$
\begin{array}{rll}
\min & u(N) & \\
\text{s.t.} & u(S) \geq 1 & \text{for all } S \in SP \\
& u \geq 0 &
\end{array}
\tag{2.15}
$$

The complementary slackness property from linear programming theory gives us the following results.

- If there exists an optimal flow $x \in R^n$ with $x_i = 0$ then $u_i = 0$ in any core element u.

■   If there exists an optimal flow $x \in R^n$ with $x_i = 1$ for all the members
     (arcs) of a given coalition $S \in SP$, then $u(S) = 1$ in every core element u.

Let v be a simple flow game with graph G. Let $u \in R^n$. By P^u we denote
the linear programming problem that we obtain when we change the objective
function in the linear programming problem that determines $v(N)$ from $p \cdot x$
to $(p - u) \cdot x$.

Proposition 2.6.1 *Let v be a simple flow game with graph G. Let x be an
optimal flow in G and let $u \in R^n$ satisfy $u(N) = v(N)$. Then $u \in C(v)$ if and
only if x is optimal for P^u with optimal value 0.*

Proof. Note that x is feasible for P^u since only the objective function has been
changed. Its value is

$$(p - u) \cdot x = \sum_{i \in N} p_i x_i - \sum_{i \in N} u_i x_i$$

$$= v(N) - \sum_{i \in N} u_i x_i.$$

Let $u \in C(v)$. Then complementary slackness tells us that $u_i > 0$ implies
$x_i = 1$. So $\sum_{i \in N} u_i x_i = u(N) = v(N)$. So $(p - u) \cdot x = 0$. Since $u(S) \geq 1$ for
every $S \in SP$ it follows that $(p - u)(S) \leq 0$ for every $S \in SP$. So the value of
P^u is 0 and x is optimal in P^u.
Let u and x be such that x is optimal for P^u with value 0. Then $(p - u)(S) \leq 0$
for all $S \in SP$. So $u(S) \geq 1$ for all $S \in SP$ and hence $u \in C(v)$. $\square$

Now we can prove that every core element of a simple flow game is generated by
an optimal solution of the dual problem of the problem that determines $v(N)$.
Consider the linear programming problem which yields $v(N)$ in a simple flow
game v.

$$\begin{aligned}
\max \quad & p \cdot x \\
\text{s.t} \quad & xI \leq \bar{1} \\
& xH = \bar{0} \\
& x \geq 0.
\end{aligned} \tag{2.16}$$

Here I is the $n \times n$-identity matrix, p and H are as defined in the proof of
theorem 2.2.5, $\bar{1}$ is the vector in R^n consisting of all 1's, and $\bar{0}$ is the vector
in R^q consisting of all 0's. Here q is the number of vertices in G. The dual

problem of 2.16 is

$$\min \quad \sum_{j \in N} y_j$$
$$\text{s.t.} \quad y_j + \sum_{k \neq s, t: j \in B(k)} z_k - \sum_{l \neq s, t: j \in E(l)} z_l \geq p_j \quad \text{for all } j \in N \qquad (2.17)$$
$$y_j \geq 0 \qquad \qquad \text{for all } j \in N.$$

Here s, t are, respectively, the source and the sink of G.

Theorem 2.6.2 *Let v be a simple flow game. Let $u \in C(v)$. Then there exists $z \in R^q$ such that (u, z) is an optimal solution for 2.17.*

Proof. Let $u \in C(v)$. Then $u \geq 0$ and $u(N) = v(N)$. So if we find a $z \in R^q$ such that the other constraints of 2.17 are satisfied, we are done. Consider the problem P^u. From proposition 2.6.1 we know that the optimal value of P^u is 0. Let $(\hat{u}, \hat{z})$ be an optimal solution to the dual problem of P^u. Then $\hat{u}(N) = 0$ and $\hat{u} \geq 0$. This implies $\hat{u} = 0$. So

$$\sum_{k \neq s, t: j \in B(k)} \hat{z}_k - \sum_{l \neq s, t: j \in E(l)} \hat{z}_l \geq p_j - u_j \text{ for all } j \in N.$$

It follows that

$$u_j + \sum_{k \neq s, t: j \in B(k)} \hat{z}_k - \sum_{l \neq s, t: j \in E(l)} \hat{z}_l \geq p_j$$

which implies that $(u, \hat{z})$ is the required optimal solution for 2.17. $\qquad \square$

Ford and Fulkerson [49] show that the minimum cuts of the graph G of a simple flow game constitute the extreme points of the set of optimal solutions of problem 2.17. Together with theorem 2.6.2 this leads to the following theorem.

Theorem 2.6.3 *Let v be a simple flow game with graph G. The extreme points of $C(v)$ are exactly the vectors u^K with*

$$u_i^K = \begin{cases} 1 & \text{if } i \in K \\ 0 & \text{otherwise} \end{cases} \qquad (2.18)$$

where K is a minimum cut of G.

Proof. Immediate from the result stated above and theorem 2.6.2. $\qquad \square$

Kalai and Zemel [67] first proved the results discussed above in a slightly different setting. Reijnierse et al. [105] consider a generalization of the simple flow games studied here. They allow directed as well as undirected arcs and they also allow, so called, *public arcs* which can be used by any coalition. A cut is defined to be a set of arcs K such that each positive flow from source to sink uses at least one arc from K. With the condition that $\bigcup_{i \in N} o(i)$ is a cut they show that theorem 2.6.3 is valid in this case also. They define the *minimum cut solution* for these simple flow games and give an axiomatic characterization of this solution concept which we will discuss here. In the following when we talk about a simple flow game we will mean as defined by Reijnierse et al. [105]. Although the player set need not correspond to the set of arcs anymore, since public arcs are also allowed, we will still identify an arc with its owner without any cause for confusion.

Definition 2.6.4 *The minimum cut solution for simple flow games assigns to each simple flow game v the set*

$$MC(v) := \{u^K \mid K \text{ is a minimum cut which does not contain a public arc.}\}$$

Here u^K is as defined in 2.18.

Definition 2.6.5 *A solution ψ on the set of simple flow games is called one person efficient if for each one player simple flow game v we have $\psi(v) = \{v(N)\}$.*

Let $x \in R^N$. The *carrier* $c(x)$ of x is defined by

$$c(x) := \{i \in N \mid x_i \neq 0\}.$$

For $S \subset N$ let x^S denote the restriction of x to R^S given by $x_i^S = x_i$ for all $i \in S$. For notational ease we will use x^{-S}, G_{-S}, and v_{-S} instead of $x^{N \setminus S}$, $G_{N \setminus S}$, and $v_{N \setminus S}$.

Definition 2.6.6 *A solution ψ on the set of simple flow games is called consistent if for each simple flow game with more than one player we have: $x \in \psi(v)$ and $S \subset c(x)$ imply $x^{-S} \in \psi(v_{-S})$.*

Definition 2.6.7 *A solution ψ on the set of simple flow games is called converse consistent if for each simple flow game v with more than one player we have: $x \in PI(v)$ and $x^{-i} \in \psi(v_{-i})$ for all $i \in c(x)$ imply $x \in \psi(v)$.*

Proposition 2.6.8 *The minimum cut solution satisfies one person efficiency, consistency, and converse consistency.*

Proof. If there is only one player then his arc forms a cut. If it is a minimum cut then it is the only minimum cut so $MC(v) = \{1\} = \{v(N)\}$. If it is not a minimum cut then $v(N) = 0$ and $MC(v) = \{0\}$.

Let v be a simple flow game. Let $x \in MC(v)$ and $S \subset c(x)$. Let K be the minimum cut in G with $x = u^K$. Then $K \setminus S$ is a cut in the graph G_{-S} of v_{-S}. Let L be a cut in G_{-S}. Then $L \cup S$ is a cut in G. Using the max flow-min cut theorem of Ford and Fulkerson we obtain

$$c(L) + c(S) = c(L \cup S) \geq v(N). \tag{2.19}$$

Also,

$$c(K \setminus S) = v(N) - c(S). \tag{2.20}$$

From 2.19 and 2.20 it follows that $K \setminus S$ is a minimum cut in G_{-S}. So $x^{-S} \in MC(v_{-S})$ and MC satisfies consistency.

Let v be a simple flow game. Let $x \in PI(v)$ with $x^{-i} \in MC(v_{-i})$ for all $i \in c(x)$. If $c(x) = \emptyset$ then $v(N) = 0$, so $0 = x \in MC(v)$.

If $c(x) = \{i\}$ then $0 \in MC(v_{-i})$, so $v(N \setminus \{i\}) = 0$. Therefore, the arc i is a cut. Because $x \in PI(v)$ we have $x_i = v(N) = 1$, and it follows that i is a minimum cut. So $x \in MC(v)$.

If $|c(x)| > 1$, take $i \in c(x)$. Because $x^{-i} \in MC(v_{-i})$, we have $x_j = 1$ for all $j \in c(x) \setminus \{i\}$. Similarly $x_i = 1$. Since $c(x) \setminus \{i\}$ is a minimum cut in G_{-i} it follows that $c(x)$ is a cut in G. From the efficiency of f we deduce that $c(x)$ is a minimum cut in G. So $x \in MC(v)$ and MC satisfies converse consistency. $\square$

Proposition 2.6.9 *Let ψ be a solution on the set of simple flow games which satisfies one person efficiency and consistency. Let η be a solution on the set of simple games which satisfies one person efficiency and converse consistency. Then $\psi(v) \subset \eta(v)$ for all simple flow games v.*

Proof. By induction on the number of players. If $|N| = 1$ then $\psi(v) = v(N) = \eta(v)$. Let $\psi(w) \subset \eta(w)$ for all simple flow games w with less that n players. Let v be an n-person simple flow game. Let $x \in \psi(v)$. Since ψ satisfies consistency we have $x^{-i} \in \psi(v_{-i}) \subset \eta(v_{-i})$ for all $i \in c(x)$. Because η satisfies converse consistency it follows that $x \in \eta(v)$. So $\psi(v) \in \eta(v)$. $\square$

Theorem 2.6.10 *The minimum cut solution is the unique solution on the set of simple flow games which satisfies one person efficiency, consistency, and converse consistency.*

Proof. From propositions 2.6.8 and 2.6.9 we know that for any solution ψ on the set of simple flow games which satisfies one person efficiency, consistency, and converse consistency $MC(v) \subset \psi(v) \subset MC(v)$ for all simple flow games v. So $\psi = MC$.   □

Two players i and j in a simple flow game are called *inseparable* if all coalitions $S \in SP$ that contain i also contain j and vice versa. In the case that all arcs are undirected and non-public Granot and Granot [56] show that the intersection of the core and the kernel equals the set of all core allocations that assign the same amount to inseparable players. In the more general case that Reijnierse et al. consider, this is not true anymore. They show that the following weaker property holds: Any element of the intersection of the core and the kernel assigns the same amount to inseparable players.

Reijnierse et al. also give conditions that guarantee that the core is a subset of the kernel. They also show that for superadditive simple flow games the bargaining set equals the core.

3

ASSIGNMENT GAMES AND PERMUTATION GAMES

3.1 INTRODUCTION

Vladimir, Wanda, and Xavier each has a house that he/she wants to sell. Yolanda and Zarik each wants to buy a house. Vladimir values his house at $100,000, Wanda values her house at $150,000, and Xavier values his house at $200,000. Vladimir's house is worth $80,000 to Yolanda and $150,000 to Zarik, Wanda's house is worth $200,000 to Yolanda and $120,000 to Zarik, and Xavier's house is worth $220,000 to Yolanda and $230,000 to Zarik. For the sake of notational shortness, we will divide all numbers by 100,000 in the following discussion. If Vladimir and Yolanda get together they will not be able to reach an agreement since Vladimir values his house at more than it is worth to Yolanda. If Vladimir and Zarik get together they can generate a profit of 1.5-1=0.5. If Wanda and Yolanda get together they can generate a profit of 2-1.5=0.5. If Wanda and Zarik get together they will not be able to reach an agreement since Wanda values her house at more than it is worth to Zarik. If Xavier and Yolanda get together then they can generate a profit of 2.2-2=0.2. If Xavier and Zarik get together then they can generate a profit of 2.3-2=0.3. If they cooperate in coalitions of more than two persons then the matching(s) between the owner of a house and a buyer that generate the most profit has to be found for each coalition. It is clear that coalitions that contain only sellers or only buyers will not generate any profit. In table 3.1, where the cooperative game arising from this situation is given, these coalitions have been left out. There Vladimir=1, Wanda=2, Xavier=3, Yolanda=4, and Zarik=4. If all five of them cooperate the highest profit is generated if Yolanda buys Wanda's house and Zarik buys Vladimir's house. This game has a non-empty core. Two core elements are (0,0,0,0.5,0.5) and (0.2,0.3,0,0.2,0.3). The first one is very favourable for the buyers, the whole profit is divided between the two of them.

53

S	$v(S)$	S	$v(S)$	S	$v(S)$
{1,4}	0	{1,2,5}	0.5	{3,4,5}	0.3
{1,5}	0.5	{1,3,4}	0.2	{1,2,3,4}	0.5
{2,4}	0.5	{1,3,5}	0.5	{1,2,3,5}	0.5
{2,5}	0	{1,4,5}	0.5	{1,2,4,5}	1
{3,4}	0.2	{2,3,4}	0.5	{1,3,4,5}	0.7
{3,5}	0.3	{2,3,5}	0.3	{2,3,4,5}	0.8
{1,2,4}	0.5	{2,4,5}	0.5	{1,2,3,4,5}	1

Table 3.1 An assignment game.

	Mon	Tue	Wed
Adams	2	4	8
Benson	10	5	2
Cooper	10	6	4

Table 3.2 The preferences schedule for a dentist appointment.

The second one is very favourable to the sellers. They get the highest payoff that they can expect from a core element.

The following example describes a situation that has much in common with the one described above, but which is not exactly the same.

Mr. Adams, Mrs. Benson, and Mr. Cooper have appointments with the dentist for next week. Mr. Adams' appointment is on Monday, Mrs. Benson's appointment is on Tuesday, and Mr. Cooper's appointment is on Wednesday. Mr. Adams would prefer to reschedule his appointment for later in the week, because he will be very busy in the beginning of the week. Mrs. Benson's toothache has evolved from a dull background pain to an acute one during the weekend, so she is very interested in being helped as soon as possible once the week starts. Mr. Cooper would like to get it over with fast, so he can concentrate on his busy schedule for the remainder of the week. The three of them discuss this situation and they decide to try to find a way to reshuffle the appointments among them to obtain a more favourable schedule. First each one of them figures out how much it is worth to him/her to be helped on a particular day. Table 3.2 summarizes this information. Using this table Mr. Adams and Mrs. Benson, for example, see that if the two of them reach an agreement without Mr. Cooper, they should change appointments which will give them a value of 14. Table 3.3 gives the cooperative game that arises from this situation. Here Mr. Adams=1, Mrs. Benson=2, and Mr. Cooper=3. The

S	$v(S)$	S	$v(S)$	S	$v(S)$	S	$v(S)$
$\{1\}$	2	$\{3\}$	4	$\{1,3\}$	18	$\{1,2,3\}$	24
$\{2\}$	5	$\{1,2\}$	14	$\{2,3\}$	9	$\emptyset$	0

Table 3.3   A permutation game.

core is the convex hull of the points (9,5,10),(8,6,10),(15,5,4),(14,6,4).

The two examples given here deal with related situations. In both cases there is a coupling of one player to another. However, there are also differences. In the first example the player set can be partitioned into two sets, and only two players from different sets of this partition will be coupled. Further, if player i is coupled to player j then j is also coupled to i. This leads to an assignment game. In the second example there is no partition of the player set and player i can be coupled to player j while player j is coupled to player k. This leads to a permutation game. In the following sections we will study these games in more detail. In section 3.2 we will introduce them formally and show that they are totally balanced. We will also discuss symmetrically pairwise bargained allocations for assignment games. In section 3.3 we will introduce generalizations of assignment games and permutation games, multi-assignment games and multi-permutation games. These games need not be balanced. We will study two classes of balanced multi-assignment games and multi-permutation games. In section 3.4 we look at economies with indivisibilities as another way to model multi-permutation situations. Finally, in section 3.5 we will consider matching situations in which the players rank the possible outcomes.

3.2   ASSIGNMENT GAMES AND PERMUTATION GAMES

Consider the following bipartite matching situation. Let M and P be two disjoint sets. For each pair $(i, j) \in M \times P$ the value of matching i and j is $a_{ij} \geq 0$. From this situation the following cooperative game v can be constructed. For each coalition $S \subset M \cup P$ the worth $v(S)$ of S is defined to be the maximum that S can achieve by making suitable pairs from its members. If $S \subset M$ or $S \subset P$ then no suitable pairs can be made and therefore $v(S) = 0$. Formally,

$v(S)$ is equal to the value of the following integer programming problem.

$$
\begin{array}{ll}
\max & \sum_{i\in M}\sum_{j\in P} a_{ij}x_{ij} \\
\text{s.t.} & \sum_{j\in P} x_{ij} \leq 1_S(i) \quad \text{for all } i \in M \\
& \sum_{i\in M} x_{ij} \leq 1_S(j) \quad \text{for all } j \in P \\
& x_{ij} \in \{0,1\} \qquad\ \text{for all } i \in M, j \in P.
\end{array}
\tag{3.1}
$$

A game as defined in 3.1 is called an *assignment game*. Assignment games were introduced by Shapley and Shubik [121]. In the example of table 3.1 the a_{ij} are defined by $a_{ij} = \max\{h_{ij} - c_i, 0\}$ where h_{ij} is the value of the house of player i for player j and c_i is the value of the house of player i for himself.

Consider the following situation. Let $N = \{1, 2\ldots, n\}$ be given. Each $i \in N$ places the value $k_{i\pi(i)}$ on a permutation $\pi \in \Pi_N$. Each coalition $S \subset N$ has the power to orchestrate a permutation π such that only the members of S are permuted, that is, $\pi(i) = i$ for all $i \in N \setminus S$. A cooperative game v can be constructed from this situation as follows. For each coalition S the worth $v(S)$ of S is defined to be the maximum of the sum of the values of all the players in S, taken over all the permutations that are feasible for S. Let Π_S be the set of all permutations $\pi \in \Pi_N$ with $\pi(i) = i$ for all $i \in N \setminus S$. Then

$$
v(S) := \max_{\pi \in \Pi_S} \sum_{i \in S} k_{i\pi(i)}.
$$

The game that we obtain in this way is called a *permutation game*. An alternative way to define $v(S)$ is as the value of the following integer programming problem.

$$
\begin{array}{ll}
\max & \sum_{i\in N}\sum_{j\in N} k_{ij}x_{ij} \\
\text{s.t.} & \sum_{j\in N} x_{ij} = 1_S(i) \quad \text{for all } i \in N \\
& \sum_{i\in N} x_{ij} = 1_S(j) \quad \text{for all } j \in N \\
& x_{ij} \in \{0,1\} \qquad\ \text{for all } i, j \in N.
\end{array}
\tag{3.2}
$$

Permutation games were introduced in Tijs et al. [136] as cost games. In the following we will explore the relationship between assignment games and permutation games.

Theorem 3.2.1 *Every assignment game is a permutation game.*

Proof. Let v be an assignment game with player set $M \cup P$. For every $i, j \in M \cup P$ define

$$
k_{ij} := \begin{cases} a_{ij} & \text{if } i \in M, j \in P \\ 0 & \text{otherwise.} \end{cases}
$$

Let w be the permutation game defined by 3.2 with k_{ij} as given above and $N = M \cup P$. Note that the number of variables in the integer programming problem which defines $v(S)$ is $|M| \times |P|$ while the number of variables in the integer programming problem which defines $w(S)$ is $(|M| \times |P|)^2$. For $S \subset M$ or $S \subset P$, $w(S) = 0 = v(S)$. Let $S \subset M \cup P$ with $S \not\subset M$ and $S \not\subset P$. Let $x \in \{0, 1\}$ be an optimal solution for 3.1. Define $\hat{x} \in \{0, 1\}^{(|M| \times |P|)^2}$ by

$$
\begin{aligned}
\hat{x}_{ij} &:= x_{ij} && \text{if } i \in M, j \in P \\
\hat{x}_{ij} &:= x_{ji} && \text{if } i \in P, j \in M \\
\hat{x}_{ii} &:= 1_S(i) - \textstyle\sum_{j \in P} x_{ij} && \text{if } i \in M \\
\hat{x}_{jj} &:= 1_S(j) - \textstyle\sum_{i \in M} x_{ij} && \text{if } j \in P \\
\hat{x}_{ij} &:= 0 && \text{in all other cases.}
\end{aligned}
$$

Then $\hat{x}$ is a, not necessarily optimal, solution for 3.2. Hence,

$$
w(S) \geq \sum_{i \in M \cup P} \sum_{j \in M \cup P} k_{ij} \hat{x}_{ij} = \sum_{i \in M} \sum_{j \in P} a_{ij} x_{ij} = v(S).
$$

On the other hand, let $z \in \{0, 1\}^{(|M| \times |P|)^2}$ be an optimal solution for 3.2. Define $\hat{z} \in \{0, 1\}^{|M| \times |P|}$ by

$$
\hat{z}_{ij} = z_{ij} \text{ for } i \in M, j \in P.
$$

Then $\hat{z}$ is a solution for 3.1. So,

$$
v(S) \geq \sum_{i \in M} \sum_{j \in P} a_{ij} \hat{z}_{ij} = \sum_{i \in M} \sum_{j \in P} k_{ij} z_{ij} = w(S).
$$

Consequently, $v = w$. $\qquad\qquad\square$

The question arises, naturally, if every permutation game is an assignment game. It is easy to see that this is not the case. A necessary condition is the existence of a partition of the player set N of the permutation game into two subsets N_1 and N_2, such that the value of a coalition S is 0 whenever $S \subset N_1$ or $S \subset N_2$. But this condition is not sufficient as the following example shows.

Example Let $N = \{1, 2, 3\}$ and let v be the permutation game with k_{ij} given in the following matrix.

$$
\begin{pmatrix}
0 & 2 & 1 \\
1 & 0 & 0 \\
2 & 0 & 0
\end{pmatrix}
$$

Then $v(i) = 0$ for all $i \in N$, $v(1,2) = v(1,3) = 3$, $v(2,3) = 0$, and $v(N) = 4$. This game fulfils the condition given above with $N_1 = \{1\}$ and $N_2 = \{2,3\}$, but it is not an assignment game as we can show as follows. If v were an assignment game then $a_{12} = a_{13} = 3$. This leads to the contradiction $4 = v(N) = \max\{a_{12}, a_{13}\} = 3$.

Theorem 3.2.2 *Assignment games and permutation games are totally balanced.*

Proof. Because of theorem 3.2.1 it is sufficient to show that permutation games are totally balanced. Since any subgame of a permutation game is also a permutation game we need only prove that a permutation game is balanced.
Let v be a permutation game. From the Birkhoff-von Neumann theorem which states that the extreme points of the set of doubly stochastic matrices are exactly the permutation matrices it follows that 3.2 is equivalent to

$$
\begin{array}{ll}
\max & \sum_{i \in N} \sum_{j \in N} k_{ij} x_{ij} \\
\text{s.t.} & \sum_{j \in N} x_{ij} = 1_S(i) \quad \text{for all } i \in N \\
& \sum_{i \in N} x_{ij} = 1_S(j) \quad \text{for all } j \in N \\
& x_{ij} \geq 0 \quad \text{for all } i, j \in N.
\end{array}
\tag{3.3}
$$

The dual problem of 3.3 is

$$
\begin{array}{ll}
\max & \sum_{i \in N} 1_S(i) y_i + \sum_{j \in N} 1_S(j) z_j \\
\text{s.t.} & y_i + z_j \geq k_{ij} \quad \text{for all } i, j \in N.
\end{array}
\tag{3.4}
$$

Let $(\hat{y}, \hat{z})$ be an optimal solution for 3.4 for $S = N$. Then

$$
\sum_{i \in N} (\hat{y}_i + \hat{z}_i) = v(N)
$$

and for all $S \in 2^N$

$$
\sum_{i \in S} (\hat{y}_i + \hat{z}_i) = \sum_{i \in N} 1_S(i) \hat{y}_i + \sum_{i \in N} 1_S(i) \hat{z}_i \geq v(S)
$$

where the inequality follows from the fact that $(\hat{y}, \hat{z})$ is a solution of problem 3.4 for all $S \in 2^N$. Therefore $u \in R^N$ defined by $u_i := \hat{y}_i + \hat{z}_i$ is an element of the core of v.   □

Because of the complimentary slackness condition we know that if $\hat{x}_{ij} = 1$ for an optimal solution $\hat{x}$ of 3.3 then $\hat{y}_i + \hat{z}_j = k_{ij}$ for any optimal solution $(\hat{y}, \hat{z})$

of 3.4. We can consider player i in a permutation game as consisting of a seller's side and a buyer's side. He sells his position (appointment day in the game of table 3.3) at a price of $\hat{y}_i$ and looks for another one to buy. In this he is maximizing $k_{ij} - \hat{y}_j$ over j. If he chooses a particular j this means that $\hat{x}_{ij} = 1$ and $\hat{y}_j + \hat{z}_i = k_{ij}$. So his payoff is the sum of his selling price plus the difference of the value of position j for him and the amount he pays for it, i.e., his payoff is $\hat{y}_i + k_{ij} - \hat{y}_j = \hat{y}_j + \hat{z}_i$ which is the amount that the core element of the proof of theorem 3.2.2 assigns to him.

Shapley and Shubik [121] show that the core of an assignment game corresponds to the set of optimal solutions of the dual problem of 3.1 for N. Balinski and Gale [5] show that the core of an assignment game can have at most $\binom{2m}{m}$ extreme points where m is the minimum of $|M|$ and $|P|$. They show that the core has the maximum number of extreme points for square games ($|M| = |P| = m$) when there are m supercompatible pairs. A pair of players $(i, j) \in M \times P$ is called *supercompatible* if for all $S \subset M \cup P$ with $i, j \in S$, $x_{ij} = 1$ in any optimal solution for 3.1.
Solymosi and Raghavan [124] give an algorithm of order $O(|M|^3|P|)$ to find the nucleolus of an assignment game. Here $|M|$ is assumed to be the minimum of $|M|$ and $|P|$.
Rochford [107] introduces *symmetrically pairwise bargained allocations* (SPB allocations) for assignment games as a way of arriving at a distribution of the profit generated by a matched pair between the two players who constitute the pair. For the assignment game of table 3.1 we can compute an SPB allocation as follows. In this game the unique optimal assignment is given by $x_{15} = 1$ and $x_{24} = 1$. Consider the core element (0,0,0,0.5,0.5). Players 1 and 5 assume that the other players, that is, 2,3,4 receive the amount allocated to them by this core element and with this information they try to reach an agreement on how to divide the 0.5 that the two of them generate. Player 1 considers what other options he has except 5. The only other player he can match up with is 4, generating a profit of 0. So the maximum threat he can level against 5 is 0. Player 5 also considers his other options. The best he can do is pair up with 3 to generate a profit of 0.3. Since 3 does not get anything in the allocation under consideration, 5 uses the complete 0.3 as a threat in the bargaining procedure with 1. Therefore, the 0.5 that the two of them generate is divided as follows

$$\begin{aligned} 1 \text{ gets} \quad & 0 + \tfrac{1}{2}(0.5 - 0 - 0.3) = 0.1 \\ 5 \text{ gets} \quad & 0.3 + \tfrac{1}{2}(0.5 - 0 - 0.3) = 0.4 \end{aligned}$$

So the new allocation has become (0.1,0,0,0.5,0.4). Players 2 and 4 start bargaining in the same way for the 0.5 that they generate assuming that the other players get the amount given to them by the allocation (0,0,0,0.5,0.5). The threat of player 2 is 0 and that of player 4 is 0.2. So

$$2 \text{ gets} \quad 0 + \tfrac{1}{2}(0.5 - 0 - 0.2) = 0.15$$
$$4 \text{ gets} \quad 0.2 + \tfrac{1}{2}(0.5 - 0 - 0.2) = 0.35$$

The new allocation is (0.1,0.15,0,0.35,0.4). This is an SPB allocation. Starting with the core element (0.2,0.3,0,0.2,0.3) we arrive at the same allocation. Straightforward verification shows that this allocation is in the intersection of the kernel and the core of the game. In the example that yielded this game this means that Yolanda buys Wanda's house for \$165,000 and that Zarik buys Vladimir's house for \$110,000.

This process can be formalized as follows. Let $\hat{x}$ be an optimal assignment for an assignment game v. Let $u \in C(v)$. For $(i^*, j^*) \in M \times P$ with $x_{ij} = 1$ we define

$$t_{i^*}(\hat{x}, u) := \max_{j \in P \setminus \{j^*\}}(v(i^*, j) - u_j)$$
$$t_{j^*}(\hat{x}, u) := \max_{i \in M \setminus \{i^*\}}(v(i, j^*) - u_i).$$

Define $\hat{t}_{i^*}(\hat{x}, u) := \max\{t_{i^*}(\hat{x}, u), 0\}$ and $\hat{t}_{j^*}(\hat{x}, u) := \max\{t_{j^*}(\hat{x}, u), 0\}$. These are the *threats* of i and j. The symmetrically bargained incomes of i^* and j^* are

$$b_{i^*}(\hat{x}, u) := \hat{t}_{i^*}(\hat{x}, u) + \tfrac{1}{2}(v(i^*, j^*) - \hat{t}_{i^*}(\hat{x}, u) - \hat{t}_{j^*}(\hat{x}, u))$$
$$b_{j^*}(\hat{x}, u) := \hat{t}_{i^*}(\hat{x}, u) + \tfrac{1}{2}(v(i^*, j^*) - \hat{t}_{i^*}(\hat{x}, u) - \hat{t}_{j^*}(\hat{x}, u)).$$

The set of *SPB allocations* SPB(v) is defined by

SPB(v) := $\{u \in C(v) | u_k = b_k(\hat{x}, u)$ for all $k \in M \cup P$ and an optimal assignment $\hat{x}\}$.

Rochford [107] proves that SPB(v) is equal to the intersection of the kernel and the core.

3.3 MULTI-ASSIGNMENT GAMES AND MULTI-PERMUTATION GAMES

In addition to their houses (without gardens) Vladimir, Wanda, and Xavier each also has a plot of land at another location from the house that they want to sell. Except for a house, Yolanda and Zarik would also like to buy a plot of land to start a garden. They value a house and plot of land together. For Yolanda the values of the several pairs are given in table 3.4. For Zarik the values are given in table 3.5. These tables should be read as follows: The number in the Vladimir row and the Wanda column represents the value of the

	Vladimir	Wanda	Xavier
Vladimir	5	2	2
Wanda	4	5	4
Xavier	3	5	3

Table 3.4   The values of the pairs for Yolanda.

	Vladimir	Wanda	Xavier
Vladimir	3	3	2
Wanda	7	3	2
Xavier	3	3	3

Table 3.5   The values of the pairs for Zarik.

pair consisting of Vladimir's house and Wanda's plot of land. A house alone or a plot of land alone does not have any value for Yolanda and Zarik. Vladimir values his house at 1 and his plot of land at 2. Wanda values her house at 2 and her plot of land at 1. Xavier values his house at 3 and his plot of land at 2. Again, we can construct a cooperative game from this situation. In this game, coalitions that consist only of sellers or only of buyers cannot achieve any profit. The value of a coalition S consisting of both sellers and buyers is equal to the maximum that S can achieve by making a suitable matching among the sellers and buyers belonging to S. The *2-assignment game* v arising from this situation is given in table 3.6. Here Vladimir=1, Wanda=2, Xavier=3, Yolanda=4, and Zarik =5. The coalitions consisting only of sellers or only of buyers have been left out. Contrary to the assignment game where only one indivisible good was involved, the multi-assignment game with more indivisible goods need not be balanced as this example shows. Take the balanced collection $\mathcal{B} = \{\{1,2,5\}, \{1,4\}, \{2,4\}, \{3\}, \{5\}\}$ with weights $\lambda_{\{3\}} = 1$, and $\lambda_S = \frac{1}{2}$ for all other coalitions $S \in \mathcal{B}$. Then

$$\sum_{S \in \mathcal{B}} \lambda_S v(S) = \frac{1}{2}(3 + 2 + 2 + 0) + 0 = 3\frac{1}{2} > 3 = v(N).$$

Hence v is not balanced.

In a straightforward way we can achieve generalizations to p-assignment games for $p > 2$. Consider a sellers-buyers market where each seller possesses one item of different types of indivisible goods, e.g., one house, one car, one television, one compact disc player, which he wants to sell. Each buyer wants to buy exactly one item of each type of good. A buyer need not buy all the goods from the same seller. Suppose there are p different types of goods. Let M be

S	$v(S)$	S	$v(S)$	S	$v(S)$
{1,4}	2	{1,2,5}	3	{1,2,3,4}	2
{1,5}	0	{1,3,4}	2	{1,2,3,5}	3
{2,4}	2	{1,3,5}	0	{1,2,4,5}	3
{2,5}	0	{1,4,5}	2	{1,3,4,5}	2
{3,4}	0	{2,3,4}	2	{2,3,4,5}	2
{3,5}	0	{2,3,5}	0	{1,2,3,4,5}	3
{1,2,4}	2	{3,4,5}	0	$\emptyset$	0

Table 3.6 A 2-assignment game.

the set of sellers and P the set of buyers. With $(i_1, i_2, \ldots, i_p) \in M^p$ we denote a bundle consisting of the good of type 1 of seller i_1, the good of type 2 of seller i_2, and so forth. For each $j \in P$ we denote the worth of a bundle $(i_1, i_2, \ldots, i_p)$ to him by $g_j(i_1, i_2, \ldots, i_p)$. For each $i \in M$ and each $q \in \{1, 2, \ldots, p\}$ we denote by $h_i(q, j)$ the worth to seller i of selling his good of type q to buyer j. This situation, where for each of the p types of goods an assignment of sellers to buyers must be made, is called a *p-assignment situation*. We can define a *p-assignment game* from this situation by taking the worth of each coalition S to be equal to the maximum that S can achieve by making a suitable matching among the sellers and buyers belonging to S. Let $a(i_1, i_2, \ldots, i_p, j) := g_j(i_1, i_2, \ldots, i_p) + h_{i_1}(1, j) + h_{i_2}(2, j) + \cdots + h_{i_p}(p, j)$ denote the value of assigning the bundle $(i_1, i_2, \ldots, i_p)$ to buyer j. Then $v(S)$ is the value of the problem

$$
\begin{array}{lll}
\max & \sum_{(i_1, i_2, \ldots, i_p, j) \in M^p \times P} a(i_1, \ldots, i_p, j) x(i_1 \ldots, i_p, j) & \\
\text{s.t.} & \sum_{(i_2, i_3, \ldots, i_p, j) \in M^{p-1} \times P} x(i, i_2, i_3, \ldots, i_p, j) \leq 1_S(i) & \text{for all } i \in M \\
& \sum_{(i_1, i_3, \ldots, i_p, j) \in M^{p-1} \times P} x(i_1, i, i_3, \ldots, i_p, j) \leq 1_S(i) & \text{for all } i \in M \\
& \sum_{(i_1, i_2, i_4, \ldots, i_p, j) \in M^{p-1} \times P} x(i_1, i_2, i, i_4, \ldots, i_p, j) \leq 1_S(i) & \text{for all } i \in M \\
& \qquad \vdots & \qquad \vdots \\
& \sum_{(i_1, \ldots, i_{p-1}, j) \in M^{p-1} \times P} x(i_1, \ldots, i_{p-1}, i, j) \leq 1_S(i) & \text{for all } i \in M \\
& \sum_{(i_1, \ldots, i_p) \in M^p} x(i_1, \ldots, i_p, j) \leq 1_S(j) & \text{for all } j \in P \\
& x(i_1, \ldots, i_p, j) \in \{0, 1\} & \text{for all } j \in P, \\
& & (i_1, \ldots, i_p) \in M^p. \\
& & \hspace{2em} (3.5)
\end{array}
$$

As we saw, these games can have an empty core. The integral conditions can not be replaced by non-negativity conditions, so the duality theorem of linear programming theory does not apply here.

	Mr. Adams			Mrs. Benson			Mr. Cooper		
	Mon	Tue	Wed	Mon	Tue	Wed	Mon	Tue	Wed
Mon	0	0	0	0	1	0	1	0	0
Tue	1	0	0	0	0	0	0	2	0
Wed	0	0	2	0	0	2	0	0	0

Table 3.7   Preferences of the players in a 2-permutation game.

Apart form appointments with the dentist, Mr. Adams, Mrs. Benson and Mr. Cooper have also made appointments with the doctor on the same days. Mr. Adams prefers to have both his appointments on Wednesday. Mrs. Benson also prefers both her appointments on Wednesday, but she will also settle for the dentist's appointment on Monday and the doctor's appointment on Tuesday. Mr. Cooper prefers both his appointments on Tuesday. In table 3.7 the values that each one of them puts on a certain combination of appointments is given. Here the rows correspond to dentist's appointments and the columns to doctor's appointments. From this situation we can construct a 2-permutation game v, where the worth $v(S)$ of a coalition S is the maximum of the sum of the values of all members of S, taken over all the permutations for which nothing changes for the players in $N \setminus S$. This game is given in table 3.8. This game is not balanced. Consider the balanced collection $\mathcal{B} = \{\{1,2\}, \{1,3\}, \{2,3\}\}$ with all weights equal to $\frac{1}{2}$. For this collection we have

$$\frac{1}{2}(v(1,2) + v(1,3) + v(2,3)) = \frac{1}{2}(2 + 3 + 4) = 4\frac{1}{2} > 4 = v(1,2,3).$$

So v has an empty core.

Again, it is quite straightforward to generalize to p-permutation games with $p > 2$. In a p-permutation game a player i places a value on a p-tuple of permutations instead of on one permutation. We can consider, for example, a machine-job situation in which each job has to be processed on p types of machines. Each producer owns one machine of each type. For producer i we denote the value he puts on using machine $\pi_1(i)$ of type 1, machine $\pi_2(i)$ of type 2, machine $\pi_3(i)$ of type 3, and so forth, by $k_i(\pi_1(i), \pi_2(i), \pi_3(i), \ldots, \pi_p(i))$. Here $\pi_1, \pi_2, \pi_3, \ldots, \pi_p$ are permutations of N, the set of producers. In the

S	$v(S)$	S	$v(S)$	S	$v(S)$	S	$v(S)$
$\{1\}$	0	$\{3\}$	0	$\{1,3\}$	3	$\{1,2,3\}$	4
$\{2\}$	0	$\{1,2\}$	2	$\{2,3\}$	4	$\emptyset$	0

Table 3.8   A 2-permutation game.

corresponding *p-permutation game* v, the worth $v(S)$ of a coalition S is the best that S can achieve by using permutations that involve only the members of S. Formally, $v(S)$ is the value of the problem

$$
\begin{aligned}
\max \quad & \sum_{(j_1,\ldots,j_p,i)\in N^{p+1}} k_i(j_1,\ldots,j_p)x(j_1,\ldots,j_p,i) \\
\text{s.t} \quad & \sum_{(j_2,j_3,\ldots,j_p,i)\in N^p} x(j,j_2,\ldots,j_p,i) = 1_S(j) && \text{for all } j \in N \\
& \sum_{(j_1,j_3,\ldots,j_p,i)\in N^p} x(j_1,j,j_3,\ldots,j_p,i) = 1_S(j) && \text{for all } j \in N
\end{aligned}
$$

$$
\begin{aligned}
& \qquad\qquad\qquad \vdots && \vdots \\[-2pt]
& \sum_{(j_1,\ldots,j_{p-1},i)\in N^p} x(j_1,j_2,\ldots,j_{p-1},j,i) = 1_S(j) && \text{for all } j \in N \\
& \sum_{(j_1,\ldots,j_p)\in N^p} x(j_1,j_2,\ldots,j_p,i) = 1_S(i) && \text{for all } i \in N \\
& x(j_1,j_2,\ldots,j_p,i) \in \{0,1\} && \text{for all } i \in N, \\
& && (j_1,\ldots,j_p) \in N^p.
\end{aligned}
$$

$$(3.6)$$

Although we have seen that p-assignment games and p-permutation games can have empty cores, it is possible to define subclasses of these games that contain only balanced games. The first class that we will consider is that of p-assignment games and p-permutation games with *additive revenues*.

Definition 3.3.1 *A p assignment game* $< M \cup P, v >$ *is said to have additive revenues if there exist real numbers* $g_j^1(i_1), g_j^2(i_2), \ldots g_j^p(i_p)$ *such that* $g_j(i_1, i_2, \ldots, i_p) = g_j^1(i_1) + \cdots + g_j^p(i_p)$ *for each* $(i_1, \ldots, i_p) \in M^p$ *and each* $j \in P$.
A p-permutation game $< N, v >$ *is said to have additive revenues if there exist real numbers* $k_i^1(j_1), k_i^2(j_2), \ldots, k_i^p(j_p)$ *such that* $k_i(j_1, j_2, \ldots, j_p) = k_i^1(j_1) + k_i^2(j_2) + \cdots + k_i^p(j_p)$ *for each* $(j_1, j_2, \ldots, j_p, i) \in N$.

A concrete situation which may give rise to a p-assignment game with additive revenues is the sellers-buyers market with p indivisible commodities that are not related to each other, e.g., a house, a car, a television. In this case it is quite plausible to assume that each commodity has a value of its own and that the value of a bundle is simply the sum of the values of the separate commodities. It is not always reasonable to assume that the revenues are additive. The commodities may be part of a big machinery which can only operate with all parts together and such that not all combinations of parts give the same performance. Then the value of a bundle really depends on the combination of commodities. ,
An example of a situation leading to a p-permutation game with additive revenues is the following. Each player has certain appointments on certain days,

for example, with the dentist, with the hairdresser, with the masseur. The value of having a certain appointment on a certain date is independent of when one has the other appointments. Then the value of a combination of appointments is simply the sum of the value of the separate appointments. If the appointments are interrelated then one cannot assume anymore that the revenues are additive.

Theorem 3.3.2 *A p-assignment (p-permutation) game with additive revenues is totally balanced.*

Proof. Let $< M \cup P, v >$ be a p-assignment game with additive revenues $g_j(i_1, \ldots, i_p) = g_j^1(i_1) + \cdots + g_j^p(i_p)$ for each $(i_1, \ldots, i_p) \in M^p$ and each $j \in P$. Define p assignment games $< M \cup P, v^1 >, < M \cup P, v^2 >, \ldots, < M \cup P, v^p >$ by taking $g_{ij}^q = g_j^q(i)$ and $h_{ij}^q = h_i(q, j)$ for each $q \in \{1, \ldots, p\}$ and each $i \in M$, $j \in P$. Then $v(S) = v^1(S) + v^2(S) + v^3(S) + \cdots + v^p(S)$ for all $S \in 2^N$. Since the games $v^1, v^2, \ldots, v^p$ are totally balanced, it follows that v is also totally balanced. For p-permutation games the proof runs similarly. $\qquad\square$

Another class of balanced p-assignment and p-permutation games is the class of these games with separable revenues. A p-assignment game with separable revenues is an appropriate model for a sellers-buyers market with p indivisible commodities where each bundle $(i_1, \ldots, i_p)$ has an intrinsic value $g(i_1, \ldots, i_p)$, and f_j denotes the value for buyer j of possessing any bundle of goods, and a seller is indifferent to whom he sells his goods, i.e., $h_i(q, j) = h_i(q)$ for all $i \in M$, $j \in P$, $q \in \{1, \ldots, p\}$.
A p-permutation game with separable revenues can be used to model a job-machine situation with p types of machines where the value of processing on a certain combination $(j_1, \ldots, j_p)$ of machines is the same for each job, namely, $k(j_1, \ldots, j_p)$. There is also an additional value which depends on the job.

Definition 3.3.3 *A p-assignment game $< M \cup P, v >$ is said to have separable revenues if there exist $f_j \geq 0$ and $a(i_1, \ldots, i_p) \geq 0$ such that $a(i_1, \ldots, i_p, j) = f_j + a(i_1, \ldots, i_p)$ for all $(i_1, i_2, \ldots, i_p) \in M^p$ and all $j \in P$.*
A p-permutation game is said to have separable revenues if there exist $k(j_1, \ldots, j_p)$ and l_i such that $k_i(j_1, \ldots, j_p) = l_i + k(j_1, \ldots, j_p)$ for all $(j_1, \ldots, j_p, i) \in N^{p+1}$.

Theorem 3.3.4 *2-assignment games and 2-permutation games with separable revenues are totally balanced.*

Proof. Let v be a 2-permutation game with separable revenues $k_i(j_1, j_2) = l_i + k(j_1, j_2)$ for each $(j_1, j_2, i) \in N^3$. Let $< N, l >$ be the additive game defined by $l(S) := \sum_{i \in S} l_i$ for each $S \in 2^N$. Let $< N, w >$ be the permutation game with $k_{ij} = k(i, j)$ for each $(i, j) \in N^2$. Then $v(S) = l(S) + w(S)$. Since l and w are totally balanced it follows that v is totally balanced.

Let $< M \cup P, v >$ be a 2-assignment game with separable revenues $a(i_1, i_2, j) = f_j + a(i_1, i_2)$ for all $(i_1, i_2) \in M^2$ and all $j \in P$. W.l.o.g. we assume that $|P| \geq |M|$. Consider the permutation game $< M, w >$ with $k_{i_1 i_2} = a(i_1, i_2)$ for all $(i_1, i_2) \in M^2$. This game is balanced. Let $y \in C(w)$. Let P' consists of $|M|$ elements of P with highest f-values and let $P'' = P \setminus P'$. Denote $\max_{i \in P''} f_i$ by f. Define $x \in R^{|M| \times |P|)}$ by $x_i = y_i + f$ for all $i \in M$, $x_i = f_i - f$ for all $i \in P'$ and $x_i = 0$ for all $i \in P''$. Then $x(N) = v(N)$ and for all $S \in 2^N$, $x(S) = y(S \cap M) + f(S \cap P') + |S \cap M|f - |S \cap P'|f \geq v(S)$. Thus $x \in C(v)$ and v is balanced. Since every subgame of a 2-assignment game with separable revenues is a 2-assignment game with separable revenues, it follows that v is totally balanced. $\qquad \square$

For $p > 2$, p-assignment games and p-permutation games with separable revenues need not be balanced, as the following example shows.

Example Let $p = 3$ and let $< M \cup P, v >$ be the 3-assignment game with separable revenues given by $M = \{1, 2, 3, 4\}$, $P = \{5, 6, 7, 8\}$ and $a(i_1, i_2, i_3, j) = f_j + a(i_1, i_2, i_3)$ where $f_j = 0$ for each $j \in P$,

$$
\begin{array}{ll}
a(1, 3, 3) = 2 & a(2, 3, 3) = 0 \\
a(1, 3, 4) = 0 & a(2, 3, 4) = 1 \\
a(1, 4, 3) = 0 & a(2, 4, 3) = 3 \\
a(1, 4, 4) = 2 & a(2, 4, 4) = 0,
\end{array}
$$

and $a(i_1, i_2, i_3) = 0$ for all combinations of (i_1, i_2, i_3) not given above. Consider the balanced collection $\mathcal{B} = \{\{2, 3, 4, 5, 6, 7\}, \{1, 3, 5, 6\}, \{1, 4, 7, 8\}, \{2, 8\}\}$ with weights $\lambda_S = \frac{1}{2}$ for all $S \in \mathcal{B}$. Then

$$
\sum_{S \in \mathcal{B}} \lambda_S v(S) = \frac{1}{2}(3 + 2 + 2 + 0) = 3\frac{1}{2} > 3 = v(M \cup P).
$$

So v is not balanced.

Let $< N, w >$ be the 3-permutation game with separable revenues given by $N = \{1, 2, 3\}$, $k_i(j_1, j_2, j_3) = l_i + k(j_1, j_2, j_3)$ where $l_i = 0$ for each $i \in N$ and the $k(j_1, j_2, j_3)$'s are given below.

$$\begin{pmatrix} 0 & 0 & 0 \\ 1 & 0 & 0 \\ 0 & 0 & 2 \end{pmatrix} \begin{pmatrix} 0 & 1 & 0 \\ 0 & 0 & 0 \\ 0 & 0 & 2 \end{pmatrix} \begin{pmatrix} 1 & 0 & 0 \\ 0 & 2 & 0 \\ 0 & 0 & 0 \end{pmatrix}$$

Here the first matrix corresponds to $j_1 = 1$, the second one to $j_1 = 2$, and the third one to $j_1 = 3$. The rows of the matrices correspond to j_2 and the columns to j_3. So $k(2, 3, 1)$ is the element in the third row and the first column of the second matrix. Consider the balanced collection $\mathcal{B} = \{\{1, 2\}, \{1, 3\}, \{2, 3\}\}$ with weights $\lambda_S = \frac{1}{2}$ for all $S \in \mathcal{B}$. Then

$$\sum_{S \in \mathcal{B}} \lambda_S v(S) = \frac{1}{2}(2 + 3 + 4) = 4\frac{1}{2} > 4 = v(1, 2, 3).$$

So v is not balanced.

In the case that a p-assignment game or p-permutation game has an empty core we still need to find a way to divide the revenues. We saw that the proof given in theorem 3.2.2 does not carry over to the case with $p > 1$ because, in general, the LP-relaxations of problems 3.5 and 3.6 do not have the same values as the integer programming problems themselves. Still we can consider allocations that we obtain from optimal solutions for the dual problems of 3.5 and 3.6. It turns out that these allocations are in the least tax core of the game. In the following we will show this for a p-assignment game v. The case of a p-permutation game goes similarly. Let $v_a(S)$ be the value of the LP-relaxation of 3.5. The dual problem of this LP-relaxation is

$$\begin{aligned} \min \quad & \sum_{i \in N} 1_S(i) y_i^1 + \cdots + \sum_{i \in N} 1_S(i) y_i^p + \sum_{i \in N} 1_S(i) z_i \\ \text{s.t.} \quad & \sum_{q=1}^{p} y_{j_q}^q + z_i \geq a(j_1, j_2, \ldots, j_p, i) \qquad \text{for all } i \in P, \\ & \qquad\qquad\qquad\qquad\qquad\qquad\qquad (j_1, \ldots, j_p) \in M^p \\ & y^1, y^2, \ldots, y^p, z \geq 0. \end{aligned}$$

$$(3.7)$$

Let $(\hat{y}^1, \hat{y}^2, \ldots, \hat{y}^p, \hat{z})$ be an optimal solution for 3.7 with $S = N$. Define $u \in R^{|M| \times |P|}$ by

$$u_i = \begin{cases} \sum_{q=1}^{p} \hat{y}_i^q & \text{for } i \in M \\ z_i & \text{for } i \in N. \end{cases} \qquad (3.8)$$

Then $u(N) = v_a(N) \geq v(N)$. Consider the allocation $\frac{v(N)}{v_a(N)} u$ in the game v. It is efficient and individual rational, but in general it will not be an element of the core. It is, however, an element of the least tax core of v. First we show that it is in the core of the multiplicative ε-tax game with $\varepsilon = \frac{v_a(N) - v(N)}{v_a(N)}$.

Lemma 3.3.5 *Let $\varepsilon = \frac{v_a(N)-v(N)}{v_a(N)}$. Then $\frac{v(N)}{v_a(N)}u \in C(v^\varepsilon)$.*

Proof. Let $S \subset N$, $S \neq N$. Because v is zero-normalized we have $v^\varepsilon(S) = (1-\varepsilon)v(S)$. So

$$v^\varepsilon(S) = \frac{v(N)}{v_a(N)}v(S) \leq \frac{v(N)}{v_a(N)}u(S).$$

The inequality follows from $v_a(S) \geq v(S)$ and the fact that $(\hat{y}^1, \hat{y}^2, \ldots, \hat{y}^p, \hat{z})$ is also feasible for the dual problem that determines $v_a(S)$. Further,

$$v^\varepsilon(N) = v(N) = \frac{v(N)}{v_a(N)}v_a(N) = \frac{v(N)}{v_a(N)}u(N).$$

So $u \in C(v^\varepsilon)$. $\qquad\square$

Remember that $\varepsilon(v)$ is the smallest ε for which the corresponding multiplicative ε-tax game has a non-empty core.

Lemma 3.3.6 *Let v be a p-assignment game. Then $\varepsilon(v) = \frac{v_a(N)-v(N)}{v_a(N)}$.*

Proof. From lemma 3.3.5 it follows that $\varepsilon(v) \leq \frac{v_a(N)-v(N)}{v_a(N)}$. Let $(1-\varepsilon(v))u \in C(v^{\varepsilon(v)})$. Then $(1-\varepsilon(v))u \geq 0$ and

$(1-\varepsilon(v))u(j_1, j_2, \ldots, j_p, i) \geq a(j_1, j_2, \ldots, j_p, i)$ for all $(j_1, j_2, \ldots, j_p, i) \in M^p \times P$.

So $(1 - \varepsilon(v))u$ is feasible for 3.7 and therefore,

$$v(N) = (1-\varepsilon(v))u(N) \geq (1-\varepsilon(v))v_a(N).$$

Consequently,

$$\varepsilon(v) \geq \frac{v_a(N) - v(N)}{v_a(N)},$$

and it follows that $\varepsilon(v) = \frac{v_a(N)-v(N)}{v_a(N)}$. $\qquad\square$

The following theorem is an immediate consequence of lemma 3.3.5 and lemma 3.3.6.

Theorem 3.3.7 *Let v be a p-assignment game. Let u be as defined in 3.8. Then $\frac{v(N)}{v_a(N)}u$ is an element of the least tax core $C(v^{\varepsilon(v)})$ of v.*

Kuipers [78] first proved theorem 3.3.7 for a slightly different version of multi-assignment games.

3.4 ECONOMIES WITH INDIVISIBILITIES

Another way to model multi-permutation situations is with *economies with indivisibilities*. An economy with indivisibilities E is an ordered tuple $< N, p, ((e_i^q)_{q \in \{1,...p\}})_{i \in N}, (\omega_i)_{i \in N}, (u_i)_{i \in N} >$ where the finite set N is the set of agents, and $p \in N$ is the number of different types of indivisible commodities in the economy. For every type of indivisible commodity q, the commodity of this type owned by agent i at the beginning is denoted by e_i^q. The amount of money that agent i possesses at the beginning is denoted by $\omega_i \in R_+$. For each agent $i \in N$, u_i is an utility function which describes the value for agent i of a bundle consisting of one indivisible commodity of each type and a certain amount of money. Let $T^q := \bigcup_{i \in N} e_i^q$ for each $q \in \{1, \ldots, p\}$. Then u_i is a function defined on $T^1 \times T^2 \times \cdots \times R_+$ with values in R. An allocation in such an economy is a distribution of the commodities and the money among the agents, such that every agent obtains exactly one commodity of each type. Such an allocation can be denoted with the aid of p-permutations, each one describing how the commodities of one type are redistributed among agents, and a vector m, describing how the money is redistributed among the agents. Thus, $(\pi^1, \pi^2, \ldots, \pi^p, m)$ is an allocation which assigns to agent i the indivisible commodities $e^1_{\pi^1(i)}, e^2_{\pi^2(i)}, \ldots, e^p_{\pi^p(i)}$ and an amount of money equal to m_i.

Every subgroup $S \subset N$ of agents can redistribute the indivisible commodities and the money available in S among its members. Let $A(S)$ be the set of *feasible allocations* for S, i.e.,

$$A(S) := \{(\pi^1, \pi^2, \ldots, \pi^p, m) \in \Pi_N^p \times R^N | \pi^q(S) = S \text{ for each } q \in \{1, \ldots, p\}$$
$$\text{and } m(S) \leq \omega(S)\}.$$

A coalition S can *improve upon* an allocation $(\pi^1, \pi^2, \ldots, \pi^p, m)$ if there exists an allocation $(\sigma^1, \sigma^2, \ldots, \sigma^p, m') \in A(S)$ such that

$$u_i(e^1_{\sigma^1(i)}, e^2_{\sigma^2(i)}, \ldots, e^p_{\sigma^p(i)}, m'_i) > u_i(e^1_{\pi^1(i)}, e^2_{\pi^2(i)}, \ldots, e^p_{\pi^p(i)}, m_i) \text{ for each } i \in S.$$

Definition 3.4.1 *The core* $C(E)$ *of an economy* E *with indivisibilities is the set of allocations that no coalition can improve upon.*

A coalition $S \subset N$ can *weakly improve upon* an allocation $(\pi^1, \pi^2, \ldots, \pi^p, m)$ if there exists an allocation $(\sigma^1, \ldots, \sigma^p, m') \in A(S)$ such that

$$u_i(e^1_{\sigma^1(i)}, \ldots, e^p_{\sigma^p(i)}, m'_i) \geq u_i(e^1_{\pi^1(i)}, \ldots, e^p_{\pi^p(i)}, m_i) \quad \text{for all } i \in S,$$
$$u_j(e^1_{\sigma^1(j)}, \ldots, e^p_{\sigma^p(j)}, m'_j) > u_j(e^1_{\pi^1(j)}, \ldots, e^p_{\pi^p(j)}, m_j) \quad \text{for at least one } j \in S.$$

Definition 3.4.2 *The strong core $SC(E)$ of an economy E with indivisibilities is the set of allocations that no coalition can weakly improve upon.*

It is evident that the strong core of an economy is contained in the core of the economy.

Definition 3.4.3 *A price equilibrium for an economy with indivisibilities E is a tuple $< p^1, \ldots, p^p, (\pi^1, \ldots, \pi^p, m) >$ where $p^q \in R_+^N$ is a price vector for the q-th commodity such that*

(a) $p_{\pi^1(i)}^1 + \cdots + p_{\pi^p(i)}^p + m_i \leq p_i^1 + \cdots + p_i^p + \omega_i$ *for all $i \in N$.*

(b) *For all $i \in N$ and $(e_{j_1}^1, \ldots, e_{j_p}^p, m_i') \in T^1 \times \cdots \times T^p \times R_+$*

if $u_i(e_{j_1}^1, \ldots, e_{j_p}^p, m_i') > u_i(e_{\pi^1(i)}^1, \ldots, e_{\pi^p(i)}^p, m_i)$

then $p_{j_1}^1 + \cdots + p_{j_p}^p + m_i' > p_i^1 + \cdots + p_i^p + \omega_i$.

If $< p^1, \ldots, p^p, (\pi^1, \ldots, \pi^p, m) >$ is a price equilibrium then the vectors $p^1, \ldots, p^p$ are called *equilibrium price vectors* and the allocation $(\pi^1, \ldots, \pi^p, m)$ is called an *equilibrium allocation*.

Theorem 3.4.4 *Let $< p^1, \ldots, p^p, (\pi^1, \ldots, \pi^p, m) >$ be a price equilibrium. Then $(\pi^1, \ldots, \pi^p, m) \in C(E)$.*

Proof. Let $S \subset N$. Suppose there exists a $(\sigma^1, \ldots, \sigma^p, m') \in A(S)$ such that

$$u_i(e_{\sigma^1(i)}^1, \ldots, e_{\sigma^p(i)}^p, m_i') > u_i(e_{\pi^1(i)}^1, \ldots, e_{\pi^p(i)}^p, m_i) \text{ for all } i \in S.$$

Then

$$p_{\sigma^1(i)}^1 + \cdots + p_{\sigma^p(i)}^p + m_i' > p_i^1 + \cdots + p_i^p + \omega_i \text{ for all } i \in S.$$

Since $\sigma^q(S) = S$ for all $q \in \{1, \ldots, p\}$ it follows that $m'(S) > \omega(S)$, contradicting $(\sigma^1, \ldots, \sigma^p, m_i') \in A(S)$. Therefore, no S can improve upon $(\pi^1, \ldots, \pi^p, m)$ and hence $(\pi^1, \ldots, \pi^p, m) \in C(E)$. $\square$.

The core of an economy with indivisibilities can be empty as the following example shows.

Example Let $N = \{1, 2, 3\}$, $p = 2$, $\omega_1 = \omega_2 = \omega_3 = 3$,

$$u_1(e_j^1, e_k^2, m_1) = \begin{cases} m_1 + 1 & \text{for } j = 2, k = 1 \\ m_1 + 2 & \text{for } j = 3, k = 3 \\ m_1 & \text{otherwise,} \end{cases}$$

$$u_2(e_j^1, e_k^2, m_2) = \begin{cases} m_2 + 1 & \text{for } j = 1, k = 2 \\ m_2 + 2 & \text{for } j = 3, k = 3 \\ m_2 & \text{otherwise,} \end{cases}$$

$$u_3(e_j^1, e_k^2, m_3) = \begin{cases} m_3 + 1 & \text{for } j = 1, k = 1 \\ m_3 + 2 & \text{for } j = 2, k = 2 \\ m_3 & \text{otherwise.} \end{cases}$$

For any allocation (π^1, π^2, m) the following holds.

$$\sum_{i \in N} u_i(e_{\pi^1(i)}^1, e_{\pi^2(i)}^2, m_i) \leq 13 \tag{3.9}$$

For an allocation $(\hat{\pi}, \hat{\sigma}, \hat{m})$ to be in the core of E the following should be satisfied.

$$\begin{array}{ll} u_1(e_{\hat{\pi}(1)}^1, e_{\hat{\sigma}(1)}^2, \hat{m}_1) + u_2(e_{\hat{\pi}(2)}^1, e_{\hat{\sigma}(2)}^2, \hat{m}_2) & \geq \quad 8 \\ u_1(e_{\hat{\pi}(1)}^1, e_{\hat{\sigma}(1)}^2, \hat{m}_1) + u_3(e_{\hat{\pi}(3)}^1, e_{\hat{\sigma}(3)}^2, \hat{m}_3) & \geq \quad 9 \\ u_2(e_{\hat{\pi}(2)}^1, e_{\hat{\sigma}(2)}^2, \hat{m}_2) + u_3(e_{\hat{\pi}(3)}^1, e_{\hat{\sigma}(3)}^2, \hat{m}_3) & \geq \quad 10 \end{array} \tag{3.10}$$

Straightforward verification shows that it is impossible for both 3.9 and 3.10 to hold simultaneously. So $C(E) = \emptyset$.

Under the assumption that u_i is continuous and non-decreasing, with respect to money, for all $i \in N$, Quinzii [103] proved that the core of the economy E is non-empty for $p = 1$. Under certain additional assumptions Quinzii proved that the core of E coincides with the set of equilibrium allocations of E.

Theorem 3.4.5 *Let* $E = < N, 2, ((e_i^q)_{q \in \{1,2\}})_{i \in N}, (\omega_i)_{i \in N}, (u_i)_{i \in N} >$ *be an economy with 2 indivisible commodities and* $u_i(e_j^1, e_k^2, m_i) = u(e_j^1, e_k^2, m_i) + f_i$ *for all* $i, j, k \in N$ *where* u *is a function from* $T^1 \times T^2 \times R_+$ *to* R *which is continuous and non-decreasing with respect to money, and where* $f_i \in R$. *Then* $C(E) \neq \emptyset$.

Proof. Define the economy $\hat{E} = < N, 1, (\hat{e}_i)_{i \in N}, (\hat{\omega}_i)_{i \in N}, (\hat{u}_i)_{i \in N} >$ by $\hat{e}_i = e_i^2$, $\hat{\omega}_i = \omega_i$ and $\hat{u}_i(\hat{e}_j, m_i) = u(e_i^1, e_j^2, m_i)$ for all $i, j \in N$. Then Quinzii's result yields $C(\hat{E}) \neq \emptyset$. Let $(\hat{\pi}, m) \in C(\hat{E})$. Then (σ, π, m) where σ is the identity permutation, is an allocation that cannot be improved upon by any coalition. Suppose, namely, that a coalition S could improve upon the allocation $(\sigma, \hat{\pi}, m)$ by means of the allocation (π^1, π^2, m'). Consider the allocation $(\pi^2(\pi^1)^{-1}, m')$ in $\hat{E}$. Let $i \in S$, then there is a j such that $i = \pi^1(j)$ and

$$
\begin{aligned}
\hat{u}_i(\hat{e}_{\pi^2(\pi^1)^{-1}(i)}, m'_i) &= \hat{u}_{\pi^1(j)}(\hat{e}_{\pi^2(j)}, m'_i) &= \\
u(e^1_{\pi^1(j)}, e^2_{\pi^2(j)}, m'_i) &> u(e^1_1, e^2_{\hat{\pi}(i)}, m_i) &= \\
\hat{u}_i(\hat{e}_{\hat{\pi}(i)}, m_i).
\end{aligned}
$$

This contradicts $(\hat{\pi}, m) \in C(\hat{E})$ and therefore it is not possible for an S to improve upon $(\sigma, \hat{\pi}, m)$. Hence $(\sigma, \hat{\pi}, m) \in C(E))$. $\qquad\square$

Note that theorem 3.4.5 is the analogue of theorem 3.3.4 in terms of economies with indivisible commodities.

3.5 ORDINAL MATCHING SITUATIONS

Two medical students Mrs. Davis and Mr. Edmonds are considering at which hospital to do their clerkships in surgery and gynaecology. They have two options, Franklin Hospital and Grant Hospital. Mrs Davis prefers to do both at Franklin Hospital. If this is not possible then she would like to do both at Grant Hospital. If this is also not possible then she will settle for doing surgery at Franklin and gynaecology at Grant. The combination that she likes the least is surgery at Grant and gynaecology at Franklin. Mr. Edmonds also has his preferences. They are given below together with the ones of Mrs. Davis

| Mrs. Davis: | FF | GG | FG | GF |
| Mr. Edmonds: | GF | FG | FF | GG |

Here F stands for Franklin Hospital and G for Grant Hospital. The first combination mentioned is the most preferred one and they continue going down in preference. In a combination the first position stands for surgery and the second for gynaecology. The hospitals also rank the students for each job. Franklin Hospital prefers Mrs. Davis for surgery and Mr. Edmonds for gynaecology and Grant Hospital has the same preferences. An independent arbiter wants to assign the students to the hospitals for each job in a reasonable way. First he considers assigning Mrs. Davis to Franklin for both jobs and Mr. Edmonds to Grant for both jobs. Pretty soon he discovers the flaw in this assignment.

Since Mr. Edmonds prefers GF to GG and Franklin Hospital prefers Mr. Edmonds to Mrs. Davis for gynaecology, the hospital and Mr. Edmonds can get together and change the assignment. So he considers assigning Mrs. Davis to Franklin for surgery and to Grant for gynaecology, and Mr. Edmonds to Grant for surgery and to Franklin for gynaecology. After some further consideration he realizes what's wrong with this. Since Mrs. Davis prefers GG to FG and Grant prefers Mrs. Davis to Mr. Edmonds for surgery they will get together and change this assignment. In the new assignment Mrs. Davis will go to Grant for both jobs and Mr. Edmonds will go to Franklin for both jobs. However, this also has a problem. Since Mr. Edmonds prefers FG to FF and Grant prefers Mr. Edmonds to Mrs. Davis for gynaecology they will change. So he considers assigning Mrs. Davis to Grant for surgery and to Franklin for gynaecology, and assigning Mr. Edmonds to Franklin for surgery and to Grant for gynaecology. But since Mrs. Davis prefers FF to GF and Franklin prefers Mrs. Davis to Mr. Edmonds for surgery they will get together to change this accordingly, resulting in Mrs. Davis being assigned to Franklin for both jobs and Mr. Edmonds to Grant for both jobs. But this is the first assignment he considered, so he has gone round in a circle without finding a solution. There is no *stable* matching in this case.

We can formalize and generalize the ideas that were introduced in the example above. Let H and G be two disjoint, finite sets. To fix the ideas we can think of H as being a set of hospitals and of G as being a set of students who have to fulfil their clerkships. Each hospital offers them the opportunity to perform p types of jobs. Each hospital offers exactly one position of each type. A student need not fulfil all the jobs at the same hospital. The students have preferences over combinations of jobs. For instance, one student may prefer to fulfil all the jobs at the same hospital, while another one prefers to work at as many hospitals as possible. The preferences of the students are defined on the set H^p. These preferences are complete, transitive and strict and therefore they can be represented by ordered lists. Typically the preference $P(g)$ of student $g \in G$ can be described as follows:

$P(g)$:   most preferred combination   ...   ...   least preferred combination.

With respect to the preferences of the hospitals two different models will be studied. In the first model each hospital $h \in H$ has separate preference relations $P^1(h), P^2(h), \ldots, P^p(H)$ defined on the set G of students. Here $P^j(h)$ describes the preference of h over the students for the j-th job. These preferences are also taken to be complete, transitive and strict. In the second case that we will consider each hospital $h \in H$ has a joint preference $P(h)$ defined on G^p.

In the first case the matching situation is called a *p-matching situation with
separate preferences*, while in the second case it is called a *p-matching situation
with joint preferences*. Note that in both models the preferences of the students
are joint preferences and that it is the difference in the type of preferences of
the hospitals that accounts for the difference in the names of the matching
situations.

Let μ be a function from G to H^p. For all $g \in G$ we denote $\mu(g)$ by
$(\mu_1(g), \mu_2(g), \ldots, \mu_p(g))$.

Definition 3.5.1 *A function μ from G to H^p is called a matching if*

$$g \neq g' \text{ implies } \mu(g) \neq \mu(g') \text{ for all } g, g' \in G.$$

Let μ be a matching. The element g of G for which $\mu_j(g) = h$ will be denoted
by $\mu_j^{-1}(h)$. By $\mu_{-j}(g)h$ we denote the element of H^p with h replacing $\mu_j(g)$
and all other coordinates equal to the corresponding ones in $\mu(g)$.

Definition 3.5.2 *A matching μ in a p-matching situation with separate preferences is called stable if there is no pair $(g, h) \in G \times H$ such that g prefers
$\mu_{-j}(g)h$ to $\mu(g)$ and h prefers g to $\mu_j^{-1}(h)$.*

For $p = 1$ Gale and Shapley [51] proved that there exist stable matchings and
described an algorithm to arrive at a stable matching. As the example that we
started with shows, a stable matching need not exist for $p > 1$. By putting an
extra condition on the preferences of the elements of G it is possible to ensure
the existence of a stable matching. This condition was introduced by Roth
[111] in a slightly different context.

Definition 3.5.3 *A preference relation P defined on a product set $X_1 \times X_2 \times
\cdots \times X_k$ is said to be responsive if there exist separate, complete, transitive,
and strict preference relations $P^1, P^2, \ldots, P^k$ on $X_1, X_2, \ldots, X_k$ such that if
two elements of $X_1 \times X_2 \times \cdots \times X_k$ differ only in the j-th place, the element
which contains the most preferred job according to P^j is preferred by P.*

Note that if a preference relation is responsive then the separate preference
relations are uniquely determined.

Theorem 3.5.4 *If in a p-matching situation with separate preferences the preferences of the elements of G are responsive, then there exists a stable matching.*

Proof. Let $P^1(g), P^2(g), \ldots, P^p(g)$ be the separate preferences of $g \in G$. Consider p 1-matching situations defined by the sets G and H and the preference relations $P^1(g), \ldots, P^p(g)$ for all $g \in G$ and $P^1(h), \ldots, P^p(h)$ for all $h \in H$. For all these situations stable matchings $\hat{\mu}_1, \hat{\mu}_2, \ldots, \hat{\mu}_p$ can be found with the Gale-Shapley algorithm. Define the matching μ by $\mu_q(g) = \hat{\mu}_q(g)$ for all $g \in G$ and all $q \in \{1, \ldots, p\}$. Then μ is stable in the p-matching situation, since the existence of a pair $(g, h) \in G \times H$ such that g prefers $\mu_{-j}(g)h$ to $\mu(g)$ and h prefers g to $\mu_j^{-1}(h)$ contradicts the stability of $\hat{\mu}_j$. $\qquad\square$

A stable matching μ between hospitals and students makes it impossible for a hospital h to prefer student g to student $\mu_j^{-1}(h)$ for the j-th job, while g prefers the combination where $\mu_j(g)$ is replaced by h to $\mu(g)$. However, it is still possible for a hospital h to prefer g for several jobs $j_1, j_2, \ldots, j_k$ to the students who are assigned to the jobs by μ, while g prefers the combination where $\mu_{j_1}(g), \ldots, \mu_{j_k}(g)$ are replaced by h to the combination originally assigned to him. Therefore, even an assignment according to a stable matching can be destabilized if the hospitals and students are allowed to change more than one of the students, respectively, jobs assigned to them. This is not the case if the matching is *strongly stable*. For $J \subset \{1, 2 \ldots, p\}$ let $\mu_{-J}(g)h$ denote that element of H^p arising from $\mu(g)$ by replacing the j-th component with h for all $j \in J$.

Definition 3.5.5 *A matching μ in a p-matching situation is called strongly stable if there is no pair (g, h) such that for some $J \subset \{1, 2, \ldots, p\}$, g prefers $\mu_{-J}(g)h$ to $\mu(g)$ and h prefers g to $\mu_j^{-1}(h)$ for all $j \in J$.*

It is obvious that every strongly stable matching is stable. When the preferences are responsive the following theorem states that the reverse statement also holds.

Theorem 3.5.6 *In a p-matching situation with separate preferences where all the preferences of the elements of G are responsive, every stable matching is strongly stable.*

Proof. Let μ be a stable matching. Suppose there exists a pair $(g, h) \in G \times H$ and a $J \subset \{1, 2, \ldots, p\}$ such that for all $j \in J$, $P^j(h)$ prefers g to $\mu_j^{-1}(h)$. It follows from the stability of μ that g prefers $\mu(g)$ to $\mu_{-j}(g)h$ for all $j \in J$. The responsiveness and transitivity of $P(g)$ leads to the conclusion that g prefers $\mu(g)$ to $\mu_{-j}(g)h$. Therefore, μ is strongly stable. $\qquad\square$

From theorem 3.5.4 and theorem 3.5.6 it follows immediately that in a p-matching situation with separate preferences where the preferences are responsive, there exists a strongly stable matching.

Let us go back to the example with which we started. Suppose that each hospital evaluates the students as a team instead of separately. Both hospitals have the same preference. They prefer the team consisting of Mrs. Davis for surgery and Mr. Edmonds for gynaecology. In the second place comes the team with Mrs Davis fulfilling both positions. In the third place is the team consisting of Mr. Edmonds for both jobs. The team they like the least is the one with Mr. Edmonds for surgery and Mrs. Davis for gynaecology. Again, an independent arbiter tries to achieve a reasonable assignment. Considering to assign Mrs. Davis for both jobs to Franklin and Mr. Edmonds for both jobs to Grant, he realizes that, since Mr. Edmonds prefers GF to GG and Franklin prefers DE to DD, this matching is not stable. The reader can verify that none of the possible matchings in this case is stable. This is an example of a 2-matching situation with joint preferences. For p-matching situations with joint preferences we have to modify the definition of a stable matching slightly. For a matching μ let $\mu^{-1}(h)$ denote the element of G^p with j-th coordinate equal to $\mu_j^{-1}(h)$ for all $j \in \{1, \ldots, p\}$ and let $\mu_{-j}^{-1}(h)g$ be the element of G^p with g replacing $\mu_j^{-1}(h)$ and all other coordinates equal to the corresponding ones in $\mu^{-1}(h)$.

Definition 3.5.7 *A matching μ in a p-matching situation with joint preferences is called stable if there does not exist a pair $(g, h) \in G \times H$ such that g prefers $\mu_{-j}(g)h$ to $\mu(g)$ and h prefers $\mu_{-j}^{-1}(h)g$ to $\mu^{-1}(h)$.*

For $p = 1$ there is no difference between the two models considered here and we already noted that in this case there is always a stable matching. The example discussed above shows that, just as in the case with separate preferences, this result cannot be extended to $p > 1$. Again, if all the preferences are responsive this is remedied.

Theorem 3.5.8 *In a p-matching situation with joint preferences where all the preferences of the elements of both G and H are responsive, there exists a stable matching.*

Proof. Let $P^1(g), \ldots, P^p(g)$ be the separate preferences of $g \in G$ and let $P^1(h), \ldots, P^p(h)$ be the separate preferences of $h \in H$. Consider p 1-matching situations defined by the sets G and H and the preferences $P^1(g), \ldots, P^p(g)$ for all $g \in G$ and $P^1(h), \ldots, P^p(h)$ for all $h \in H$. With the Gale-Shapley algorithm we can find stable matchings $\hat{\mu}_1, \ldots, \hat{\mu}_p$ for all these situations. Define the matching μ by $\mu_q(g) = \hat{\mu}_q(g)$ for all $g \in G$ and all $q \in \{1, \ldots, p\}$. Suppose that there exists a pair $(g, h) \in G \times H$ such that g prefers $\mu_{-j}(g)h$ to $\mu(g)$ and h prefers $\mu_{-j}^{-1}(h)g$ to $\mu^{-1}(h)$. Then in the 1-matching situation with preferences $p^j(g)$ and $P^j(h)$, g prefers h to $\hat{\mu}_j(g)$ and h prefers g to $\hat{\mu}_j^{-1}(h)$ contradicting the stability of $\hat{\mu}_j$. Therefore, such a pair (g, h) cannot exist and μ is stable. □

For p-matching situations with joint preferences we can also consider strongly stable matchings. Again, it is clear that a strongly stable matching is necessarily stable. But, contrary to the case of p-matching situations with separate preferences, in this case the responsiveness of the preferences does not guarantee that a stable matching will be strongly stable. Indeed, even the existence of a strongly stable matching is not guaranteed by the responsiveness of the preferences as the following example shows.

Example Consider the 2-matching situation with $G = \{g_1, g_2\}$, $H = \{h_1, h_2\}$ and where the preferences are given by

$$
\begin{array}{llllll}
P(g_i): & h_1 h_2 & h_1 h_1 & h_2 h_2 & h_2 h_1 & \text{for } i \in \{1, 2\} \\
P(h_1): & g_1 g_2 & g_2 g_2 & g_1 g_1 & g_2 g_1 & \\
P(h_2): & g_1 g_2 & & & g_2 g_1 &
\end{array}
$$

where the open places in $P(h_2)$ are not important. These preferences are responsive. The separate preferences are

$$
\begin{array}{llll}
P^1(g_i): & h_1 & h_2 & \text{for } i \in \{1, 2\} \\
P^2(g_i): & h_2 & h_1 & \text{for } i \in \{1, 2\} \\
P^1(h_k): & g_1 & g_2 & \text{for } k \in \{1, 2\} \\
P^2(h_k): & g_2 & g_1 & \text{for } k \in \{1, 2\}
\end{array}
$$

The only stable matching here is the matching μ with $\mu(g_1) = (h_1, h_1)$ and $\mu(g_2) = (h_2, h_2)$, but μ is not strongly stable since g_2 prefers (h_1, h_1) to (h_2, h_2) and h_1 prefers (g_2, g_2) to (g_1, g_1). Since every strongly stable matching is stable, it follows that there does not exist a strongly stable matching in this situation.

The following stronger condition on the preferences in a p-matching situation with joint preferences will guarantee the existence of a strongly stable matching.

Definition 3.5.9 *A preference relation P defined on a set X^k is said to be strongly responsive if there exists a complete, transitive, and strict preference relation $\hat{P}$ on X such that if two elements of X^k differ only in the j-th place, the element with the most preferred j-th coordinate according to $\hat{P}$ is preferred by P.*

Note that a strongly responsive preference relation on X^k defines a unique preference relation on X.

A student $g \in G$ has a strongly responsive preference if and only if he can rank the hospitals in such a way that if he prefers hospital h to hospital h' then he always prefers to fulfil a certain job at h rather than at h', regardless of where he fulfils the other jobs. Similarly we can interpret strong responsiveness for hospitals.

It is clear that a strongly responsive preference is responsive.

Theorem 3.5.10 *In a p-matching situation with joint preferences where all the preferences are strongly responsive there exists a strongly stable matching.*

Proof. For all $g \in G$ let $\hat{P}(g)$ be the preference relation defined on H by the strongly responsive preference relation $P(g)$ of g on H^p and for all $h \in H$ let $\hat{P}(h)$ be the preference relation defined on G by the strongly responsive preference relation $P(h)$ of h on G^p. Consider the 1-matching situation defined by G, H and the preference relations $\hat{P}(g)$ for all $g \in G$ and $\hat{P}(h)$ for all $h \in H$. Let $\hat{\mu}$ be a stable matching for this situation. Define a matching μ in the p-matching situation by $\mu_j(g) = \mu(g)$ for all $j \in \{1, \ldots, p\}$ and all $g \in G$. Let $(g, h) \in G \times H$ and $J \subset \{1, \ldots, p\}$ be such that g prefers $\mu_{-J}(g)H$ to $\mu(g)$. This implies that in the 1-matching situation g prefers h to $\hat{\mu}(g)$. From the stability of $\hat{\mu}$ it follows that h prefers $\hat{\mu}^{-1}(h)$ to g. The strong responsiveness and the transitivity of $P(h)$ yield that in the p-matching situation h prefers $\mu^{-1}(h)$ to $\mu_{-J}^{-1}(h)g$. It follows that μ is strongly stable. $\square$

The result stated in theorem 3.5.10 will still be true if we weaken the condition and ask for only one of the sets G and H to have members with strongly responsive preferences while the other has members with responsive preferences.

4

SEQUENCING GAMES AND GENERALIZATIONS

4.1 INTRODUCTION

Four mathematics professors, Mrs. Hewitt, Mr. Isaacs, Mrs. Jones, and Mr. Kent have handed in work to be copied to the mathematics department's copy room. Only one copy machine is functioning so they have to wait for their turn. Mrs. Hewitt was the first to hand in her job which is a referee report that she has completed. She needs to have copies made before she can send it to the editor. Mr. Isaacs, the second one to hand in his job, needs copies of exams that he has to give the next day in class, so he is in a hurry. Mrs. Jones, who handed in third, wants them to make copies of a proposal that she has to send out to apply for a grant. The last one to hand in his job, Mr. Kent, needs copies of a paper that he wants to submit to a journal. By talking to each other they get a notion that there must be a better order to make copies than the order in which they have handed in the work. Since they are mathematicians they decide to analyze the situation systematically. Each one of them writes down how much time his job will take and also how much each additional hour will cost her/him. Mrs. Hewitt's job will take 1 hour and her per hour cost is 1, Mr. Isaacs' job will take 5 hours and his per hour costs are 10, Mrs. Jones' job will take 6 hours and her per hour costs are 8, Mr. Kent's job will take 2 hours and his per hour costs are 3. They calculate that if they do not change the order their total costs will be

$$1 \cdot 1 + 10 \cdot 6 + 8 \cdot 12 + 3 \cdot 14 = 1 + 60 + 96 + 42 = 199.$$

After some deliberation they reach the conclusion that they save the most if they require that first Mr. Isaacs' job be done, then Mr. Kent's, then Mrs. Jones, and finally Mrs. Hewitt's. With this order their total costs will be

$$10 \cdot 5 + 3 \cdot 7 + 8 \cdot 13 + 1 \cdot 14 = 50 + 21 + 104 + 14 = 189.$$

So they have made a costs savings of 10. They will need a way to divide this among them. Mrs. Hewitt, for example, will have to be compensated for moving so far back. One way to obtain an allocation is to consider the following way of reaching the optimal order. First Mrs. Hewitt and Mr. Isaacs switch places. This generates a cost savings of 5 which is divided equally among them. The order is now: Mr. Isaacs, Mrs. Hewitt, Mrs. Jones, Mr. Kent. Then Mrs. Hewitt and Mrs. Jones switch places. This generates a cost savings of 2 which is divided equally among them and yields the order: Mr.Isaacs, Mrs. Jones, Mrs. Hewitt, Mr. Kent. Then Mrs. Hewitt and Mr. Kent switch places. This generates a cost savings of 1 which is divided equally among them. The order is now: Mr. Isaacs, Mrs. Jones, Mr. Kent, Mrs. Hewitt. Finally, Mrs. Jones and Mr. Kent switch places to reach the optimal order, generating a cost savings of 2 which is divided equally among them. This way of dividing the cost savings yields the allocation (4,2.5,2,1.5).

In section 4.2 we will formalize the ideas introduced here. We will consider the EGS-rule for sequencing situations, which yields the allocation given above. We will discuss two axiomatic characterizations of this rule. We will also study a generalization of the EGS-rule, the split core. We will introduce sequencing games and show that they are convex. Nice expressions for the Shapley-value and the τ-value of a sequencing game will be derived. We will introduce the head-tail core and the head-tail nucleolus and we will show that the head-tail nucleolus of a sequencing game is equal to the allocation given by the EGS-rule. In section 4.3 we will look at two generalizations of sequencing games, σ-pairing games and σ component additive games. We will define a generalization of the EGS-rule for σ-component additive games, the β-rule. Finally, we will discuss results on the bargaining set and nucleolus of Γ-component additive games, a further generalization of σ-component additive games.

4.2 SEQUENCING GAMES

Consider the following *sequencing situation*. A set of customers N is standing in a queue in front of a counter waiting to be served. The original position of each customer is given by a permutation σ of N. For $i \in N$, $\sigma(i) = j$ means that i has the j-th position in the queue. Each customer i has a certain *service time* of $s_i > 0$ units of time and a *cost function* $c_i : R_+ \to R$. The cost of customer i depends on his waiting time plus his service time. If the sum of these two equals t then his cost is $c_i(t)$. In the following we assume that c_i is linear for all i. So there exist $\alpha_i, \beta_i \in R$ such that $c_i(t) = \alpha_i t + \beta_i$. The total cost of N if served according to σ is given by

$$C_\sigma := c_1\Big(\sum_{i\in P(\sigma,1)} s_i + s_1 \Big) + c_2\Big(\sum_{i\in P(\sigma,2)} s_i + s_2 \Big) + \cdots + c_n\Big(\sum_{i\in P(\sigma,n)} s_i + s_n \Big).$$

By rearranging its members N can decrease this cost. Each rearrangement corresponds to a permutation $\pi \in \Pi_N$. The gain achieved by this rearrangement is equal to $C_\sigma - C_\pi$. Note that the β_i's do not play a role in the computation of the cost savings induced by a rearrangement. Therefore, they are not included in the following definition of a sequencing situation.

Definition 4.2.1 *A sequencing situation consists of a finite set $N = \{1, 2 \ldots, n\}$ and a ordered triple $(\sigma; \alpha; s)$ where $\sigma \in \Pi_N$, $\alpha \in R^n$, and $s \in R_+^n$.*

Whenever no confusion is possible about the identity of the set N it will not be mentioned explicitly.

To find an optimal permutation, i.e., a permutation that minimizes the total cost of N the *urgency indices* of the members of N are useful.

Definition 4.2.2 *In a sequencing situation $(\sigma; \alpha; s)$ the urgency index u_i of $i \in N$ is given by $u_i := \alpha_i / s_i$.*

For a certain permutation τ of N let $i, j \in N$ be neighbours with i standing in front of j, i.e., $\tau(j) = \tau(i) + 1$. If i and j switch positions the total cost changes with an amount $\alpha_j s_i - \alpha_i s_j$ regardless of the positions of i and j in the queue. If $u_i < u_j$ this amount will be positive and the total cost will decrease. If $u_i > u_j$ this amount will be negative and the total cost will increase. If $u_i = u_j$ this amount will be zero and the total cost stays the same. This leads to the following proposition first proven by Smith [122].

Proposition 4.2.3 *Let $(\sigma; \alpha; s)$ be a sequencing situation. Then $C_\pi = \min_{\tau \in \Pi_N} C_\tau$ if and only if $u_{\pi^{-1}(1)} \geq u_{\pi^{-1}(2)} \geq \cdots \geq u_{\pi^{-1}(n)}$.*

Proof. Let τ be a permutation which does not satisfy the condition on the right hand side. Then there exist $i, j \in N$ with $\tau(j) = \tau(i) + 1$ and $u_i < u_j$. By switching i and j the cost will decrease and it follows that C_τ is not minimal. Let ρ be a permutation which does satisfy the right hand side condition. It is possible to go from any permutation τ of N to ρ by successive switches of neighbours i, j with i in front of j and $u_j \geq u_i$. Such switches will either decrease the cost of N or leave it unchanged. It follows that $C_\tau \geq C_\rho$. So $C_\rho = \min_{\pi \in \Pi_N} C_\pi$. $\square$

If the customers decide to save money by rearranging their positions they will need a way to divide the cost savings that they generate.

Definition 4.2.4 *A division rule for sequencing situations is a function f which assigns to every sequencing situation consisting of a set N and an ordered triplet $(\sigma; \alpha; s)$ a payoff vector $f(\sigma; \alpha; s) = (f_1(\sigma; \alpha; s), \ldots, f_n(\sigma; \alpha; s))$ such that*

(a) $f_i(\sigma; \alpha; s) \geq 0$ *for each $i \in N$. (individual rationality)*

(b) $\sum_{i \in N} f_i(\sigma; \alpha; s) = C_\sigma - \min_{\tau \in \Pi_N} C_\tau$. *(efficiency)*

An example of a division rule is the *equal division rule* which divides the cost savings equally among the members of N. A drawback of this method is that it does not distinguish between customers who actually contribute to the savings and those who do not. The rule that we are going to introduce does not have this disadvantage. Let us define the *gain* g_{ij} that players i and j, who are standing next to each other with i in front of j, can achieve by switching positions. This gain is equal to the difference of the sums of the costs of i and j before and after they change places. If $u_i \geq u_j$, then i and j cannot gain anything by switching positions. If $u_j > u_i$ then i and j can gain $\alpha_j s_i - \alpha_i s_j$. Therefore,

$$g_{ij} := (\alpha_j s_i - \alpha_i s_j)_+ = \max\{\alpha_j s_i - \alpha_i s_j, 0\}.$$

Now we are ready to introduce our division rule.

Definition 4.2.5 *The Equal Gain Splitting rule (EGS-rule) is the division rule for sequencing situations which assigns to every sequencing situation $(\sigma; \alpha; s)$ the vector $(EGS_1(\sigma; \alpha; s), \ldots, EGS_n(\sigma; \alpha; s))$ defined by*

$$EGS_i(\sigma; \alpha; s) := \frac{1}{2} \sum_{k \in P(\sigma, i)} g_{ki} + \frac{1}{2} \sum_{j : i \in P(\sigma, j)} g_{ij} \text{ for each } i \in N.$$

We saw that in a sequencing situation an optimal arrangement can be reached by successive switches of neighbours who are not standing in decreasing order of urgency index. The *EGS*-rule divides the gain that two neighbours generate by switching equally between them. The reader can easily check that the *EGS*-rule is indeed a division rule. The allocation that we computed for the example in the introduction is the one given by the *EGS*-rule. We can characterize the *EGS*-rule by the following three properties.

Definition 4.2.6 *Let $(\sigma; \alpha; s)$ be a sequencing situation. Then $i \in N$ is said to be a dummy in $(\sigma; \alpha; s)$ if $\sigma(j) > \sigma(i)$ implies $u_j \leq u_i$ and $\sigma(k) < \sigma(i)$ implies $u_k \geq u_i$ for all $j, k \in N$.*

It is not necessary for a dummy to switch with anybody to arrive at an optimal order. A switch with a dummy will not generate any cost savings.

Definition 4.2.7 *Two sequencing situations $(\sigma; \alpha; s)$ and $(\tau; \alpha; s)$ are called i-equivalent if $P(\sigma, i) = P(\tau, i)$.*

Note that the completion time (waiting time plus service time) and therefore also the cost of i will be the same in two i-equivalent sequencing situations.

Definition 4.2.8 *Let $(\sigma; \alpha; s)$ be a sequencing situation and let $i, j \in N$ with $|\sigma(i) - \sigma(j)| = 1$. Then $(\tau; \alpha; s)$ with $\tau(i) = \sigma(j)$, $\tau(j) = \sigma(i)$ and $\tau(k) = \sigma(k)$ for all $k \in N \setminus \{i, j\}$ is called the ij-inverse of $(\sigma; \alpha; s)$.*

The only difference between a sequencing situation and its ij-inverse is that i and j have switched places.

Definition 4.2.9 *A division rule f for sequencing situations is said to satisfy the dummy property if $f_i(\sigma; \alpha; s) = 0$ whenever i is a dummy in $(\sigma; \alpha; s)$.*

Definition 4.2.10 *A division rule f for sequencing situations is said to satisfy the equivalence property if $f_i(\sigma; \alpha; s) = f_i(\tau; \alpha; s)$ for each $i \in N$ and each pair of i-equivalent sequencing situations $(\sigma; \alpha; s)$ and $(\tau; \alpha; s)$.*

Definition 4.2.11 *A division rule f for sequencing situations is said to satisfy the switch property if*

$$f_i(\tau; \alpha; s) - f_i(\sigma; \alpha; s) = f_j(\tau; \alpha; s) - f_j(\sigma; \alpha; s)$$

for all $i, j \in N$ and all pairs of sequencing situations $(\sigma; \alpha; s)$ and $(\tau; \alpha; s)$ where $(\tau; \alpha; s)$ is the ij-inverse of $(\sigma; \alpha; s)$.

Theorem 4.2.12 *The EGS-rule is the unique division rule for sequencing situations that satisfies the dummy property, the equivalence property, and the switch property.*

Proof. Let $(\sigma; \alpha; s)$ be a sequencing situation in which i is a dummy. Then $g_{ki} = 0$ for all $k \in P(\sigma, i)$ and $g_{ij} = 0$ for all j with $i \in P(\sigma, j)$. From the definition of the EGS-rule it follows that $EGS_i(\sigma; \alpha; s) = 0$, so the EGS-rule possesses the dummy property.

Let $(\sigma; \alpha; s)$ and $(\tau; \alpha; s)$ be two i-equivalent sequencing situations. Then $P(\sigma, i) = P(\tau, i)$ and hence

$$\{j \in N | i \in P(\sigma, j)\} = \{j \in N | i \in P(\tau, j)\}.$$

From the definition of the EGS-rule it follows that $EGS_i(\sigma; \alpha; s) = EGS_i(\tau; \alpha; s)$ so the EGS-rule possesses the equivalence property.

Let $(\sigma; \alpha; s)$ be a sequencing situation and $(\tau; \alpha; s)$ its ij-inverse. W.l.o.g. we assume that $\sigma(i) < \sigma(j)$. Then

$$EGS_i(\tau; \alpha; s) - EGS_i(\sigma; \alpha; s) = \frac{1}{2}g_{ji} - \frac{1}{2}g_{ij} = EGS_j(\tau; \alpha; s) - EGS_j(\sigma; \alpha; s).$$

Thus, the EGS-rule possesses the switch property.

Suppose f is a division rule for sequencing situations which possesses these three properties. Let the set $\mathcal{M}_\sigma$ of misplaced pairs of neighbours in a sequencing situation $(\sigma; \alpha; s)$ be defined by

$$\mathcal{M}_\sigma := \{(i, j) | \sigma(i) = \sigma(j) + 1, \; u_i > u_j\}.$$

We will prove by induction on the cardinality of $\mathcal{M}_\sigma$ that f is equal to the EGS-rule. Let $(\sigma; \alpha; s)$ be a sequencing situation with $\mathcal{M}_\sigma = \emptyset$, then everybody is a dummy and it follows that $f(\sigma; \alpha; s) = (0, 0, \ldots, 0) = EGS(\sigma; \alpha; s)$. Suppose that $f(\sigma; \alpha; s) = EGS(\sigma; \alpha; s)$ for all sequencing situations $(\sigma; \alpha; s)$ with $|\mathcal{M}_\sigma| \leq m$, where $m \geq 0$. Let $(\tau; \alpha; s)$ be such that $|\mathcal{M}_\tau| = m + 1$. Then there exist a sequencing situation $(\sigma; \alpha; s)$ and a pair $(k, l) \in \mathcal{M}_\tau$ such that $\sigma(i) = \tau(i)$ for all $i \neq \{k, l\}$ and $\sigma(k) = \tau(l)$, $\sigma(l) = \tau(k)$. Since $\mathcal{M}_\sigma = \mathcal{M}_\tau \setminus \{k, l\}$, the equivalence property and the induction assumption yield

$$f_i(\tau; \alpha; s) = f_i(\sigma; \alpha; s) = EGS_i(\sigma; \alpha; s) = EGS_i(\tau; \alpha; s)$$

for all $i \notin \{k, l\}$. Further, $C_\tau - C_\sigma = g_{kl}$ and the efficiency, the switch property and the induction assumption yield

$$f_k(\tau; \alpha; s) = f_k(\sigma; \alpha; s) + \frac{1}{2}g_{lk} = EGS_k(\sigma; \alpha; s) + \frac{1}{2}g_{lk} = EGS_k(\tau; \alpha; s).$$

Similarly, we obtain $f_l(\tau; \alpha; s) = EGS_l(\tau; \alpha; s)$. So $f(\tau; \alpha; s) = EGS(\tau; \alpha; s)$.
□

The theory of cooperative games can be used to further analyze sequencing situations. To do this we will define *sequencing games*. In a sequencing situation not only the group N but every subgroup S of N can decrease its costs by rearranging its members. In this process of rearrangement it is not allowed that any member of S jumps over non-members of S. Thus, if $i, j \in S$ and there is a $k \in N \setminus S$ standing between i and j, then i and j are not allowed to change positions. Before we define the characteristic function v of a sequencing game formally, we will introduce some concepts that will be helpful in the definition and subsequent analysis.

Definition 4.2.13 *Let $(\sigma; \alpha; s)$ be a sequencing situation. Then $T \subset N$ is called a connected coalition if for all $i, j \in T$ and $k \in N$, $\sigma(i) < \sigma(k) < \sigma(j)$ implies $k \in T$.*

In a connected coalition there are no non-members of the coalition standing between the members and therefore all rearrangements of its members are allowed. From proposition 4.2.3 it follows that an optimal way to rearrange the members of a connected coalition is in order of decreasing urgency index. This can be done by switching neighbours who are not standing in the right order. Every time this is done with two neighbours i and j the cost savings generated is equal to g_{ij}. This leads to the following definition of the worth $v(T)$ of a connected coalition.

$$v(T) := \sum_{i \in T} \sum_{k \in P(\sigma, i) \cap T} g_{ki}. \tag{4.1}$$

Definition 4.2.14 *Let $S \subset N$ be a non-connected coalition in a sequencing situation $(\sigma; \alpha; s)$. Then $T \subset N$ is said to be a component of S if the following three conditions hold.*

(a) $T \subset S$.

S	$v(S)$	S	$v(S)$	S	$v(S)$	S	$v(S)$
{1}	0	{1,2}	5	{2,4}	0	{1,3,4}	2
{2}	0	{1,3}	0	{3,4}	2	{2,3,4}	2
{3}	0	{1,4}	0	{1,2,3}	7	{1,2,3,4}	10
{4}	0	{2,3}	0	{1,2,4}	5	$\emptyset$	0

Table 4.1 The sequencing game derived from the example in the introduction.

(b) T *is connected.*

(c) *For every* $i \in N$, $T \cup \{i\}$ *is not connected.*

The components of a non-connected coalition S in a sequencing situation $(\sigma; \alpha; s)$ form a partition of S which is denoted by S/σ. Maximal cost savings are achieved by S when the members in all its components are rearranged in order of decreasing urgency index. The total cost savings of S is the sum of the cost savings of all its components. Formally, the worth $v(S)$ of a non-connected coalition S is given by

$$v(S) := \sum_{T \in S/\sigma} v(T). \qquad (4.2)$$

Definition 4.2.15 *A cooperative game* v *is a sequencing game if there exists a sequencing situation* $(\sigma; \alpha; s)$ *such that for a connected coalition* T *the definition of* $v(T)$ *is given by 4.1 and for a non-connected coalition* S *the definition of* $v(S)$ *is given by 4.2.*

The sequencing game corresponding to the sequencing situation of the example in the introduction is given in table 4.1 with Mrs. Hewitt=1, Mr. Isaacs=2, Mrs. Jones=3, and Mr. Kent=4. It is easy to check that this game is convex. The following theorem states that this is true for sequencing games in general.

Theorem 4.2.16 *Sequencing games are convex games.*

Proof. Let v be a sequencing game arising from the sequencing situation $(\sigma; \alpha; s)$. Let $S_1 \subset S_2 \subset N \setminus \{i\}$. Then there exist $T_1, U_1 \in S_1/\sigma \cup \{\emptyset\}$ and $T_2, U_2 \in S_2/\sigma \cup \{\emptyset\}$ with $T_1 \subset T_2$ and $U_1 \subset U_2$ such that

$$v(S_p \cup \{i\}) - v(S_p) = \sum_{k \in T_p} g_{ki} + \sum_{j \in U_p} g_{ij} + \sum_{k \in T_p, j \in U_p} g_{kj} \text{ for } p \in \{1,2\}.$$

From the non-negativity of the g_{ij}'s it follows that

$$v(S_1 \cup \{i\}) - v(S_1) \leq v(S_2 \cup \{i\}) - v(S_2),$$

so v is convex. □

It is evident that not all convex games are sequencing games since sequencing games are zero-normalized and convex games need not be zero-normalized. Actually, not even all zero-normalized convex games are sequencing games. In the following proposition a characterization of the 3-person zero-normalized convex games which are sequencing games is given.

Proposition 4.2.17 *A 3-person zero-normalized convex game v is a sequencing game if and only if*

(a) *at least one 2-person coalition has worth 0,*

(b) $v(N) = v(S) + v(T)$ *with $|S| = |T| = 2$ implies $v(S) = 0$ or $v(T) = 0$,*

(c) $v(S) = 0$ *for all coalitions S with $|S| = 2$ implies $v(N) = 0$.*

Proof. Let $N = \{1, 2, 3\}$ and suppose that the game v arises from the sequencing situation $(\sigma; \alpha; s)$. W.l.o.g. we assume that $\sigma(1) = 1$, $\sigma(2) = 2$, $\sigma(3) = 3$. Then $v(1, 3) = 0$ by definition. Suppose $v(N) = v(S) + v(T)$ with $|S| = |T| = 2$ and $v(S) > 0$, $v(T) > 0$. Then $S = \{1, 2\}$ and $T = \{2, 3\}$ and $u_2 > u_1$ and $u_3 > u_2$. So

$$v(N) = g_{12} + g_{23} + g_{13} > g_{12} + g_{23} = v(S) + v(T).$$

This contradicts $v(N) = v(S) + v(T)$. Consequently, either $v(S) = 0$ or $v(T) = 0$. Suppose $v(S) = 0$ for all S with $|S| = 2$, then $u_1 \geq u_2 \geq u_3$ and hence $v(N) = 0$ as well. This completes the proof of the "only if" part.
For the "if" part, note that there are three possible cases: either only one 2-person coalition or two 2-person coalitions or all three 2-person coalitions have worth zero. In the last case v is the game with $v(S) = 0$ for all $S \in 2^N$. This is a sequencing game arising from a sequencing situation where the original order already minimizes the costs, i.e., a sequencing situation $(\sigma; \alpha; s)$ with $u_{\sigma^{-1}(1)} \geq u_{\sigma^{-1}(2)} \geq u_{\sigma^{-1}(3)}$.
W.l.o.g. we assume that in the first case $v(1, 3) = 0$ and in the second case $v(1, 3) = v(1, 2) = 0$. Let $(\rho; \gamma; t)$ and $(\rho; \delta; w)$ be sequencing situations with $\rho(1) = 1$, $\rho(2) = 2$, $\rho(3) = 3$,

$$\gamma = (0, v(1, 2), v(1, 2, 3) - v(1, 2) - v(2, 3)),$$

$$t = (1, \frac{v(2,3) + v(1,2)}{v(1,2,3) - v(1,2) - v(2,3)}, 1),$$

$$\delta = (v(2,3), 0, v(2,3)), \text{ and } w = (\frac{v(1,2,3)}{v(2,3)}, 1, 1).$$

Straightforward verification shows that in the first case v is a sequencing game arising from $(\rho; \gamma; t)$ and in the second case v is the sequencing game from (ρ, δ, w). This completes the proof. $\square$

Because sequencing games are convex games they are totally balanced. Although the EGS-rule was introduced in definition 4.2.5 directly for sequencing situations without considering sequencing games, it turns out that the allocation it generates for a sequencing situation is an element of the core of the corresponding sequencing game.

Theorem 4.2.18 *Let v be the sequencing game corresponding to the sequencing situation $(\sigma; \alpha; s)$. Then $EGS(\sigma; \alpha; s) \in C(v)$.*

Proof. Since the EGS-rule satisfies efficiency we know that

$$\sum_{i \in N} EGS_i(\sigma; \alpha; s) = v(N).$$

Let $S \in 2^N$, then

$$\sum_{i \in S} EGS_i(\sigma; \alpha; s) = \sum_{i \in S} \frac{1}{2}(\sum_{k \in P(\sigma,i)} g_{ki} + \sum_{j:i \in P(\sigma,j)} g_{ij}) \geq$$

$$\sum_{i \in S} \frac{1}{2}(\sum_{k \in P(\sigma,i) \cap S} g_{ki} + \sum_{j \in S: i \in P(\sigma,j)} g_{ij}) = \sum_{i \in S} \sum_{k \in P(\sigma,i) \cap S} g_{ki} \geq v(S).$$

Thus, $EGS(\sigma; \alpha; s) \in C(v)$. $\square$

Hamers et al. [62] give a generalization of the EGS-rule which they call the *split core*. They consider all allocations obtained by dividing the gain generated by the switch of player i and j between i and j, not necessarily equally. Formally, a *gain splitting rule* GS^λ is given by

$$GS_i^\lambda(\sigma; \alpha; s) := \sum_{k \in P(\sigma,i)} \lambda_{ki} g_{ki} + \sum_{j:i \in P(\sigma,j)} (1 - \lambda_{ij}) g_{ij} \text{ for all } i \in N.$$

Here $0 \leq \lambda_{ij} \leq 1$ for all $i, j \in N$. Note that GS^λ is equal to the *EGS*-rule if $\lambda_{ij} = \frac{1}{2}$ for all $i, j \in N$.

Example If in the example of the introduction we take

$$\lambda_{12} = \frac{1}{3} \quad \lambda_{23} = \frac{5}{6}$$
$$\lambda_{13} = \frac{3}{4} \quad \lambda_{24} = \frac{2}{3}$$
$$\lambda_{14} = \frac{1}{2} \quad \lambda_{34} = \frac{1}{4}$$

we obtain

$$GS_1^\lambda = \frac{2}{3} \cdot 5 + \frac{1}{4} \cdot 2 + \frac{1}{2} \cdot 1 = 4\frac{1}{3}$$
$$GS_2^\lambda = \frac{1}{3} \cdot 5 + \frac{1}{6} \cdot 0 + \frac{1}{3} \cdot 0 = 1\frac{2}{3}$$
$$GS_3^\lambda = \frac{3}{4} \cdot 2 + \frac{5}{6} \cdot 0 + \frac{3}{4} \cdot 2 = 3$$
$$GS_4^\lambda = \frac{1}{2} \cdot 1 + \frac{2}{3} \cdot 0 + \frac{1}{4} \cdot 2 = 1.$$

The split core contains all allocations generated by gain splitting rules.

Definition 4.2.19 *The split core $SPC(\sigma; \alpha; s)$ of a sequencing situation $(\sigma; \alpha; s)$ is defined by*

$$SPC(\sigma; \alpha; s) := \{GS^\lambda | 0 \leq \lambda_{ij} \leq 1\}.$$

It is easy to verify that all the gain splitting rules satisfy efficiency, individual rationality, the dummy property and the equivalence property. Obviously, they cannot all satisfy the switch property, only the *EGS*-rule satisfies this among them. However, each one satisfies the following property which is a modification of the switch property.

Definition 4.2.20 *A division rule for sequencing situations f is said to satisfy the monotonicity property if*

$$(f_i(\tau; \alpha; s) - f_i(\sigma; \alpha; s))(f_j(\tau; \alpha; s) - f_j(\sigma; \alpha; s)) \geq 0$$

for all $i, j \in N$ and all pairs of sequencing situations $(\sigma; \alpha; s)$ and $(\tau; \alpha; s)$ with $(\tau; \alpha; s)$ the ij-inverse of $(\sigma; \alpha; s)$.

Hamers et al. [62] define the following monotonicity property for solution concepts, which may contain more than one element, and use it together with efficiency and the dummy property to characterize the split core.

Definition 4.2.21 *A solution concept F for sequencing situations is said to satisfy the monotonicity property if for all $i, j \in N$ and all pairs of sequencing situations $(\sigma; \alpha; s)$ and $(\tau; \alpha; s)$ with $(\tau; \alpha; s)$ the ij-inverse of $(\sigma; \alpha; s)$ the* *following holds: for all $x \in F(\sigma; \alpha; s)$ there exists a $y \in F(\tau; \alpha; s)$ such that*

$$x_k = y_k \text{ for all } k \in N \setminus \{i, j\} \text{ and } (x_i - y_i)(x_j - y_j) \geq 0.$$

They show that all solution concepts that satisfy efficiency, the dummy property, and monotonicity are contained in the split core.

Similarly to the proof for the *EGS*-rule it can be shown that all gain splitting rules generate allocations that are in the core of the corresponding sequencing games. Hence, the split core is a subset of the core. Since sequencing games are convex games, we know that the marginal vectors are the extreme points of the core. To find the extreme points of the split core, Hamers et al. [62] consider gain splitting rules which are based on permutations. Let $\pi \in \Pi_N$. Define $\lambda(\pi)$ by

$$\lambda_{ij}(\pi) := \begin{cases} 0 & \text{if } \pi(i) > \pi(j) \\ 1 & \text{if } \pi(i) < \pi(j) \end{cases}$$

Then $GS^{\lambda(\pi)}$ is a gain splitting rule which is based on the permutation π.

Example Consider the example of the introduction. Let π be the permutation which arranges the players as follows: 3421. That is, $\pi(1) = 4$, $\pi(2) = 3$, $\pi(3) = 1$, and $\pi(4) = 2$. Then

$$\begin{array}{llll}
\lambda_{12}(\pi) = 0 & \lambda_{21}(\pi) = 1 & \lambda_{31}(\pi) = 1 & \lambda_{41}(\pi) = 1 \\
\lambda_{13}(\pi) = 0 & \lambda_{23}(\pi) = 0 & \lambda_{32}(\pi) = 1 & \lambda_{42}(\pi) = 1 \\
\lambda_{14}(\pi) = 0 & \lambda_{24}(\pi) = 0 & \lambda_{34}(\pi) = 1 & \lambda_{43}(\pi) = 0
\end{array}$$

and

$$\begin{array}{lllll}
GS_1^{\lambda(\pi)} = & (1 - \lambda_{12})g_{12} & + & (1 - \lambda_{13})g_{13} & + & (1 - \lambda_{14})g_{14} = 8 \\
GS_2^{\lambda(\pi)} = & \lambda_{12}g_{12} & + & (1 - \lambda_{23})g_{23} & + & (1 - \lambda_{24})g_{24} = 0 \\
GS_3^{\lambda(\pi)} = & \lambda_{13}g_{13} & + & \lambda_{23}g_{23} & + & (1 - \lambda_{34})g_{34} = 0 \\
GS_4^{\lambda(\pi)} = & \lambda_{14}g_{14} & + & \lambda_{24}g_{24} & + & \lambda_{34}g_{34} = 2
\end{array}$$

So $GS^{\lambda(\pi)} = (8, 0, 0, 2)$.

We will show that the extreme points of the split core are exactly the allocations generated by gain splitting rules based on permutations. For each sequencing situation $(\sigma; \alpha; s)$ we define the *relaxed sequencing game* w by

$$w(S) := \sum_{i \in S} \sum_{k \in P(\sigma, i) \cap S} g_{ki} \text{ for all } S \subset N.$$

S	$w(S)$	S	$w(S)$	S	$w(S)$	S	$w(S)$
{1}	0	{1,2}	5	{2,4}	0	{1,3,4}	5
{2}	0	{1,3}	2	{3,4}	2	{2,3,4}	2
{3}	0	{1,4}	1	{1,2,3}	7	{1,2,3,4}	10
{4}	0	{2,3}	0	{1,2,4}	6	∅	0

Table 4.2 The relaxed sequencing game derived from the example in the introduction.

Let v be the sequencing game corresponding to $(\sigma; \alpha; s)$. Note that $v(S) = w(S)$ if S is a connected coalition, and that $v(S) \leq w(S)$ if S is not connected.

Example The relaxed sequencing game w corresponding to the example in the introduction is given in table 4.2.

Just like sequencing games relaxed sequencing games turn out to be convex games.

Theorem 4.2.22 *Relaxed sequencing games are convex games.*

Proof. Let $(\sigma; \alpha; s)$ be a sequencing situation and let w be the corresponding relaxed sequencing game. Let $T \subset S \subset N \setminus \{i\}$. Then

$$w(S \cup \{i\}) - w(S) = \sum_{k \in S \cap P(\sigma,i)} g_{ki} + \sum_{j \in S: i \in P(\sigma,j)} g_{ij}$$

$$\geq \sum_{k \in T \cap P(\sigma,i)} g_{ki} + \sum_{j \in S: i \in P(\sigma,j)} g_{ij}$$

$$= w(T \cup \{i\}) - w(T)$$

So w is convex. $\qquad\qquad\square$

Since w is convex we know that the extreme points of $C(w)$ are the marginal vectors $m^{\pi}(w)$. The following theorem shows that each marginal vector is equal to an allocation generated by a gain splitting rule based on a permutation. For convenience's sake we introduce the following notation for the set of followers of $i \in N$.

$$F(\sigma, i) := \{j \in N | \sigma(j) > \sigma(i)\}.$$

Theorem 4.2.23 *Let* $(\sigma; \alpha; s)$ *be a sequencing situation and let* w *be the corresponding relaxed sequencing game. Then* $m^{\pi}(w) = GS^{\lambda(\pi)}(\sigma; \alpha; s)$ *for all* $\pi \in \Pi_N$.

Proof. For $i \in N$ we have

$$
\begin{aligned}
m_i^{\pi}(w) &= w(P(\pi, i) \cup \{i\}) - w(P(\pi, i)) \\
&= \sum_{k \in P(\pi,i) \cap P(\sigma,i)} g_{ki} + \sum_{j \in P(\pi,i) \cap F(\sigma,i)} g_{ij} \\
&= \sum_{k \in P(\sigma,i)} \lambda_{ki}(\pi) g_{ki} + \sum_{j \in F(\sigma,i)} (1 - \lambda_{ij}(\pi)) g_{ij} \\
&= GS_i^{\lambda(\pi)}(\sigma; \alpha; s).
\end{aligned}
$$

So $m^{\pi}(w) = GS^{\lambda(\pi)}(\sigma; \alpha; s)$. $\qquad\qquad\qquad\qquad\qquad\qquad\square$

The reader is invited to work out the statement of theorem 4.2.23 for the example discussed in this and the previous section.
It is clear that $C(w) \subset C(v)$. In the next theorem we show that $C(w)$ is equal to the split core $SPC(\sigma; \alpha; s)$.

Theorem 4.2.24 *Let* $(\sigma; \alpha; s)$ *be a sequencing situation and let* w *be the corresponding relaxed sequencing game. Then* $C(w) = SPC(\sigma; \alpha; s)$.

Proof. Since w is a convex game we know that $C(w)$ is the convex hull of the marginal vectors $m^{\pi}(v)$ and because of theorem 4.2.23 it is also the convex hull of the allocations generated by gain splitting rules which are based on permutations. Because $SPC(\sigma; \alpha; s)$ is a convex set it follows that $C(w) \subset SPC(\sigma; \alpha; s)$.
Let $x \in SPC(\sigma; \alpha; s)$. Then there is a λ such that $x = GS^{\lambda}(\sigma; \alpha; s)$. For all $S \subset N$ we have

$$
\begin{aligned}
w(S) &= \sum_{i \in S} \sum_{k \in S \cap P(\sigma,i)} g_{ki} \\
&= \sum_{i \in S} \frac{1}{2} \Big(\sum_{k \in S \cap P(\sigma,i)} g_{ki} + \sum_{j \in S \cap F(\sigma,i)} g_{ij} \Big) \\
&\leq \sum_{i \in S} GS_i^{\lambda}(\sigma; \alpha; s) = x(S)
\end{aligned}
$$

where the inequality becomes an equality when $S = N$.
So $SPC(\sigma; \alpha; s) \subset C(w)$ and the proof is completed. □

The following theorem follows immediately from theorems 4.2.23 and 4.2.24.

Theorem 4.2.25 *The extreme points of the split core are exactly the allocations generated by gain splitting rules that are based on permutations.*

As the following theorem shows, the EGS-rule lies in the barycenter of the split core. The proof uses the fact that half of the $n!$ gain splitting rules which are permutation based, assign all the profit that i and j generate to i, while the other half assigns all of it to j.

Theorem 4.2.26 *Let $(\sigma; \alpha; s)$ be a sequencing situation. Then*

$$EGS(\sigma; \alpha; s) = \frac{1}{n!} \sum_{\pi \in \Pi_N} GS^{\lambda(\pi)}(\sigma; \alpha; s).$$

Proof. For $\pi \in \Pi_N$, let $\pi^c \in \Pi_N$ be the permutation corresponding to the order that is reverse to that of π. That is, for all $i, j \in N$,

$$\pi^c(i) < \pi^c(j) \text{ if and only if } \pi(i) > \pi(j).$$

Then $\lambda_{ij}(\pi) + \lambda_{ij}(\pi^c) = 1$ for all $i, j \in N$. It follows that

$$GS^{\lambda(\pi)}(\sigma; \alpha; s) + GS^{\lambda(\pi^c)}(\sigma; \alpha; s) = 2EGS(\sigma; \alpha; s),$$

and consequently,

$$
\begin{aligned}
EGS(\sigma; \alpha; s) &= \frac{1}{n!} \sum_{\pi \in \Pi_N} EGS(\sigma; \alpha; s) \\
&= \frac{1}{n!} \sum_{\pi \in \Pi_N} (\frac{1}{2} GS^{\lambda(\pi)}(\sigma; \alpha; s) + \frac{1}{2} GS^{\lambda(\pi^c)}(\sigma; \alpha; s)) \\
&= \frac{1}{n!} \sum_{\pi \in \Pi_N} GS^{\lambda(\pi)}(\sigma; \alpha; s)
\end{aligned}
$$

and the proof is completed. □

It follows immediately from theorems 4.2.23 and 4.2.26 that the allocation generated by the EGS-rule for the sequencing situation $(\sigma; \alpha; s)$ and the Shapley-value of the corresponding relaxed sequencing game coincide.

We have paid ample attention to the EGS-rule and other gain splitting rules. These rules can be defined for sequencing situations without looking at the corresponding sequencing games. Still, they turned out to have nice game theoretic properties. In the following we will consider game theoretic solution concepts for sequencing games. We will see that the Shapley-value and the τ-value of a sequencing game can be described in terms of the parameters of the corresponding sequencing situation. The following theorem says that the Shapley-value divides the gain two players can generate, equally among them and the players standing between them.

Theorem 4.2.27 *Let v be a sequencing game corresponding to the sequencing situation $(\sigma; \alpha; s)$ Then the Shapley-value $\phi(v)$ of v is given by*

$$\phi_i(v) = \sum_{\sigma(k) \le \sigma(i) \le \sigma(j)} \frac{g_{kj}}{\sigma(j) - \sigma(k) + 1} \ \text{for all } i \in N.$$

Proof. For each $i, j \in N$ with $\sigma(i) < \sigma(j)$ define the game v_{ij} by

$$v_{ij} = \begin{cases} g_{ij} & \text{if } \{l|\sigma(i) \le \sigma(l) \le \sigma(j)\} \subset S \\ 0 & \text{otherwise.} \end{cases}$$

Let T be a connected coalition, then $v_{ij}(T) = g_{ij}$ if and only if $\{i, j\} \subset T$. For a non-connected coalition S

$$v_{ij}(S) = \sum_{T \in S/\sigma} v_{ij}(T).$$

It follows that for a connected coalition T

$$\sum_{\sigma(k) < \sigma(j)} v_{kj}(T) = \sum_{j \in T} \sum_{k \in P(\sigma, j) \cap T} g_{kj} = v(T)$$

and for a non-connected coalition S

$$\sum_{\sigma(k) < \sigma(j)} v_{kj}(S) = \sum_{\sigma(k) < \sigma(j)} \sum_{T \in S/\sigma} v(T) = \sum_{T \in S/\sigma} v(T) = v(S).$$

Thus,

$$v = \sum_{\sigma(k) < \sigma(j)} v_{kj}.$$

Because of the efficiency, symmetry and dummy player property of the Shapley-value (see section 1.5) it follows that

$$\phi_i(v_{kj}) = \begin{cases} \frac{g_{kj}}{\sigma(j) - \sigma(k) + 1} & \text{if } i \in \{l | \sigma(k) \leq \sigma(l) \leq \sigma(j)\} \\ 0 & \text{otherwise.} \end{cases}$$

From the additivity of the Shapley-value it follows that

$$\phi_i(v) = \sum_{\sigma(k) < \sigma(j)} \phi_i(v_{kj}) = \sum_{\sigma(k) \leq \sigma(i) \leq \sigma(j)} \frac{g_{kj}}{\sigma(j) - \sigma(k) + 1}$$

which completes the proof. □

The τ-value of a sequencing game also divides the gain generated by two players among them and the players standing between them, but differently from the Shapley-value.

Theorem 4.2.28 *Let v be a sequencing game corresponding to the sequencing situation $(\sigma; \alpha; s)$. Then the τ-value $\tau(v)$ of v is given by*

$$\tau_i(v) = \sum_{\sigma(k) \leq \sigma(i) \leq \sigma(j)} g_{kj} \lambda \text{ for all } i \in N$$

where

$$\lambda = \frac{\sum_{j \in N} \sum_{k \in P(\sigma, j)} g_{kj}}{\sum_{j \in N} \sum_{\sigma(k) \leq \sigma(l)} g_{kl}}$$

Proof. Since v is convex it is also semiconvex and theorem 1.9.3 tells us that $\tau(v) = \lambda M^v + (1 - \lambda)\underline{v}$. Since $v(i) = 0$ for all $i \in N$ this reduces to $\tau(v) = \lambda M^v$. Also,

$$M_i^v = v(N) - v(N \setminus \{i\}) = \sum_{\sigma(k) \leq \sigma(i) \leq \sigma(j)} g_{kj}.$$

Because the τ-value is efficient it follows that

$$\lambda = \frac{v(N)}{\sum_{j \in N} M_j^v} = \frac{\sum_{j \in N} \sum_{k \in P(\sigma, j)} g_{kj}}{\sum_{j \in N} \sum_{\sigma(k) \leq \sigma(j) \leq \sigma(l)} g_{kl}}$$

which completes the proof. □

Example The Shapley-value of the sequencing game of table 4.1 is

$(3\frac{5}{12}, 3\frac{5}{12}, 1\frac{11}{12}, 1\frac{1}{4})$, the τ-value is $(3\frac{1}{3}, 3\frac{1}{3}, 2\frac{1}{12}, 1\frac{1}{4})$, and the nucleolus is $(3\frac{1}{4}, 3\frac{1}{4}, 2, 1\frac{1}{2})$.

Several authors, cf. Myerson [85], Owen [95], Faigle [44], have studied games where not all coalitions are feasible. For such a *restricted game* the solution concepts are defined while taking account of only the feasible coalitions. In the following we will consider a variation on the core of a sequencing game, introduced in [20], the so called *head-tail core* which is derived from a restriction of the game. In the definition of the head-tail core we consider only certain coalitions, the heads or tails. For player $i \in N$ in the sequencing situation $(\sigma; \alpha; s)$ the *head* $\overline{P(i)}$ is equal to the set of predecessors of i together with i, i.e., $\overline{P(i)} = P(\sigma, i) \cup \{i\}$, and the *tail* is equal to the set $F(\sigma, i)$ of followers of i. Define $\mathcal{H}$ by

$$\mathcal{H} := \{S \subset N | \text{ there is an } i \in N \text{ such that } S = \overline{P(i)} \text{ or } S = F(\sigma, i)\}.$$

Definition 4.2.29 *The head-tail core $C^{\mathcal{H}}(v)$ of the sequencing game v corresponding to the sequencing situation $(\sigma; \alpha; s)$ is given by*

$$C^{\mathcal{H}}(v) := \{x \in R^N | x(N) = v(N), \ x(S) \geq v(S) \text{ for all } S \in \mathcal{H}\}.$$

The head-tail core of v is equal to the core of the restricted game of v where the restriction is given by the set $\mathcal{H}$. It is obvious that $C(v) \subset C^{\mathcal{H}}(v)$.

Example For the sequencing game of table 4.1 we have

$$\mathcal{H} = \{\{1\}, \{2, 3, 4\}, \{1, 2\}, \{3, 4\}, \{1, 2, 3\}, \{4\}, \{1, 2, 3, 4\}\}.$$

The head-tail core is defined by

$$
\begin{array}{rcl}
x_1 + x_2 + x_3 + x_4 & = & 10 \\
x_1 & \geq & 0 \\
x_2 + x_3 + x_4 & \geq & 2 \\
x_1 + x_2 & \geq & 5 \\
x_3 + x_4 & \geq & 2 \\
x_1 + x_2 + x_3 & \geq & 7 \\
x_4 & \geq & 0
\end{array}
$$

Note that this system is equivalent to the system

$$
\begin{array}{rclll}
10 & \leq & x_1 + x_2 + x_3 + x_4 & \leq & 10 \\
0 & \leq & x_1 & \leq & 8 \\
5 & \leq & x_1 + x_2 & \leq & 8 \\
7 & \leq & x_1 + x_2 + x_3 & \leq & 10
\end{array}
$$

The head-tail core is the convex hull of the extreme points $(0, 5, 2, 3)$, $(8, -3, 2, 3)$, $(0, 8, -1, 3)$, $(8, 0, -1, 3)$, $(0, 5, 5, 0)$, $(8, -3, 5, 0)$, $(0, 8, 2, 0)$, $(8, 0, 2, 0)$.

In the following we will give a characterization of the extreme points of the head-tail core. Recall that the dual game v^* of v is defined by

$$v^*(S) = v(N) - v(N \setminus S) \text{ for all } S \subset N.$$

Note that $v^*(N) = v(N)$ and that $v(S) \leq v^*(S)$ for all $S \subset N$. From the definition of $C^{\mathcal{H}}(v)$ it follows that

$$C^{\mathcal{H}} = \{x \in R^N | v(\overline{P(i)}) \leq x(\overline{P(i)}) \leq v^*(\overline{P(i)}) \text{ for all } i \in N\}.$$

Consider the linear transformation which assigns to each $x \in C^{\mathcal{H}}$ the vector $y \in R^N$ with

$$y_i = x(\overline{P(\sigma^{-1}(i))}) \text{ for all } i \in N.$$

The matrix L of this linear transformation is the $n \times n$ lower triangular matrix with ones on and below the main diagonal. L defines a 1-1 correspondence between $C^{\mathcal{H}}$ and

$$L(C^{\mathcal{H}}) = \{y \in R^N | v(\overline{P(i)}) \leq y_i \leq v^*(\overline{P(i)}) \text{ for all } i \in N\}.$$

Therefore, there exists a 1-1 correspondence between the extreme points of both sets. Each extreme point of $L(C^{\mathcal{H}})$ is given by a system of n equations of the form

$$y_i = v_i \text{ with } v_i \in \{v(\overline{P(i)}), v^*(\overline{P(i)})\}$$

for all $i \in N$. For each $J \subset N$ we define the extreme point y^J of $L(C^{\mathcal{H}})$ by

$$y_i^J = \begin{cases} v(\overline{P(i)}) & \text{for } i \in J \\ v^*(\overline{P(i)}) & \text{for } i \in N \setminus J \end{cases}$$

Let $x^J = L^{-1}(y^J)$. Then the following theorem is an immediate consequence of the preceding discussion.

Theorem 4.2.30 *The extreme points of $C^{\mathcal{H}}(v)$ are the vectors x^J.*

Example For the game of table 4.1 we have

$$
\begin{array}{llllllll}
L(0,5,2,3) &=& (0,5,7,10) &=& y^{\{1,2,3\}} &=& y^{\{1,2,3,4\}} \\
L(8,-3,2,3) &=& (8,5,7,10) &=& y^{\{2,3\}} &=& y^{\{2,3,4\}} \\
L(0,8,-1,3) &=& (0,8,7,10) &=& y^{\{1,3\}} &=& y^{\{1,3,4\}} \\
L(8,0,-1,3) &=& (8,8,7,10) &=& y^{\{3\}} &=& y^{\{3,4\}} \\
L(0,5,5,0) &=& (0,5,10,10) &=& y^{\{1,2\}} &=& y^{\{1,2,4\}} \\
L(8,-3,5,0) &=& (8,5,10,10) &=& y^{\{2\}} &=& y^{\{2,4\}} \\
L(0,8,2,0) &=& (0,8,10,10) &=& y^{\{1\}} &=& y^{\{1,4\}} \\
L(8,0,2,0) &=& (8,8,10,10) &=& y^{\emptyset} &=& y^{\{4\}}
\end{array}
$$

The number of extreme points is equal to $2^{|I|}$ where $I = \{i \in N | v(\overline{P(i)}) \neq v^*(\overline{P(i)})\}$. In the generic case this will be equal to 2^{n-1} since $v^*(N) = v(N)$. We will show that the allocation given by the EGS-rule lies in the barycenter of the head-tail core. Consider the extreme points x^N and $x^{\emptyset}$. For each $i \in N$ we have

$$
\begin{aligned}
L^{-1}(y^N)_i &=& v(\overline{P(i)}) - v(P(\sigma,i)) \\
&=& v(P(\sigma,i) \cup \{i\}) - v(P(\sigma,i)) \\
&=& m_i^{\sigma}(v)
\end{aligned}
$$

and

$$
\begin{aligned}
L^{-1}(y^{\emptyset})_i &=& v^*(\overline{P(i)}) - v^*(P(\sigma,i)) \\
&=& v^*(F(\sigma,i) \cup \{i\}) - v(F(\sigma,i)) \\
&=& m_i^{\sigma^{-1}}(v)
\end{aligned}
$$

where σ^{-1} is the permutation corresponding to the reverse order of the one given by σ. So $x^N = m^{\sigma}(v)$ and $x^{\emptyset} = m^{\sigma^{-1}}(v)$.

Lemma 4.2.31 *Let v be the sequencing game corresponding to the sequencing situation $(\sigma; \alpha; s)$. Then*

$$
EGS(\sigma; \alpha; s) = \frac{1}{2}(m^{\sigma}(v) + m^{\sigma^{-1}}(v)).
$$

Proof. For each $i \in N$ we have

$$
EGS_i(\sigma; \alpha; s) = \frac{1}{2}\left(\sum_{k \in P(\sigma,i)} g_{ki} + \sum_{j \in F(\sigma,i)} g_{ij} \right)
$$

$$= \frac{1}{2}(v(P(\sigma,i) \cup \{i\}) - v(P(\sigma,i)))$$

$$+ \frac{1}{2}(v(F(\sigma,i) \cup \{i\}) - v(F(\sigma,i)))$$

$$= \frac{1}{2}(m_i^\sigma(v) + m_i^{\sigma^{-1}}(v)).$$

$\square$

Theorem 4.2.32 *Let v be the sequencing game corresponding to the sequencing situation $(\sigma; \alpha; s)$. Then*

$$EGS(\sigma; \alpha; s) = \frac{1}{2^n} \sum_{J \subset N} x^J.$$

Proof. For all $i \in N$ and all $J \subset N$ we have

$$y_i^J + y_i^{N \setminus J} = v(\overline{P(i)}) + v^*(\overline{P(i)}).$$

So for all $J \subset N$,

$$x^J + x^{N \setminus J} = L^{-1}(y^J + y^{N \setminus J}) = L^{-1}(y^\emptyset + y^N) = x^\emptyset + x^N.$$

Consequently,

$$\frac{1}{2^n} \sum_{J \subset N} x^J = \frac{1}{2^{n+1}} \sum_{J \subset N} (x^J + x^{N \setminus J})$$

$$= \frac{1}{2^{n+1}} 2^n (x^N + x^\emptyset)$$

$$= \frac{1}{2}(x^N + x^\emptyset)$$

$$= EGS(\sigma; \alpha; s).$$

Here the last equality follows from lemma 4.2.31. $\square$

In defining the head-tail core $C^\mathcal{H}(v)$ we considered only the inequalities that had to do with coalitions $S \in \mathcal{H}$. We can modify the definition of the (pre)-nucleolus in the same way to obtain the *head-tail (pre)-nucleolus* or *$\mathcal{H}$-(pre)-nucleolus*. This means that we will only consider the excesses of the coalitions that belong to $\mathcal{H}$. The $\mathcal{H}$-(pre)-nucleolus consists of those elements of the (pre)-imputation set that lexicographically minimizes the vector of these excesses. Because $C^\mathcal{H} \neq \emptyset$ it follows that the $\mathcal{H}$-pre-nucleolus is also non-empty and that it is a subset of $C^{\mathcal{H}}$. So the $\mathcal{H}$-pre-nucleolus and the $\mathcal{H}$-nucleolus coincide. We will show that the allocation generated by the EGS-rule is the unique element of the $\mathcal{H}$-nucleolus. First we will show that the allocation generated by the EGS-rule is the unique one that satisfies efficiency and the *fair head-tail split property* to be defined below.

Definition 4.2.33 *Let v be a sequencing game corresponding to the sequencing situation $(\sigma; \alpha; s)$. A vector $x \in R^n$ is said to satisfy the fair head-tail split property if*

$$x(S) - v(S) = x(N \setminus S) - v(N \setminus S) = \frac{1}{2}\Delta_S \ for \ all \ S \in \mathcal{H}.$$

Here $\Delta_S = v(N) - v(S) - v(N \setminus S)$.

Note that for any vector $x \in R^N$ which satisfies the fair head-tail property we have $e(S, x) = e(N \setminus S, x)$ for all $S \in \mathcal{H}$.

Theorem 4.2.34 *Let v be the sequencing game corresponding to the sequencing situation $(\sigma; \alpha; s)$. Then $EGS(\sigma; \alpha; s)$ is the unique vector in R^N that satisfies efficiency and the fair head-tail split property.*

Proof. Straightforward verification shows that $EGS(\sigma; \alpha; s)$ satisfies the fair head-tail split property. Suppose $x \in R^N$ satisfies efficiency and the fair head-tail split property. Then for $k \in N \setminus \{\sigma^{-1}(n)\}$

$$x(\overline{P(k)}) - v(\overline{P(k)}) = x(F(\sigma, k)) - v(F(\sigma, k)) \qquad (4.3)$$

and

$$x(P(\sigma, k)) - v(P(\sigma, k)) = x(F(\sigma, k) \cup \{k\}) - v(F(\sigma, k) \cup \{k\}). \qquad (4.4)$$

By subtracting 4.4 from 4.3 we obtain

$$x_k - v(\overline{P(k)} + v(P(\sigma, k)) = -x_k - v(F(\sigma, k)) + v(F(\sigma, k) \cup \{k\}).$$

So

$$x_k = \frac{1}{2}(v(\overline{P(k)}) - v(P(\sigma, k)) + v(F(\sigma, k) \cup \{k\}) - v(F(\sigma, k)) = EGS_k(\sigma; \alpha; s).$$

Because of efficiency we also have $x_n = EGS_n(\sigma; \alpha; s)$. $\square$

Theorem 4.2.35 *Let v be the sequencing game corresponding to the sequencing situation $(\sigma; \alpha; s)$. Then $EGS(\sigma; \alpha; s)$ is the unique element of the $\mathcal{H}$-nucleolus.*

Proof. Let x be an element of the $\mathcal{H}$-nucleolus and therefore also of the $\mathcal{H}$-prenucleolus. Suppose there is a $k \in N$ and an $\varepsilon > 0$ such that

$$e(\overline{P(k)}, x) - e(F(\sigma, k), x) > \varepsilon.$$

Define $y \in PI(v)$ by $y_i = x_i$ for all $i \in N \setminus \{k, \sigma^{-1}(\sigma(k) + 1)\}$, $y_k = x_k + \varepsilon$ and $y_{\sigma^{-1}(\sigma(k)+1)} = x_{\sigma^{-1}(\sigma(k)+1)} - \varepsilon$. Then

$$e(S, y) = e(S, x) \text{ for all } S \in \mathcal{H} \setminus \{\overline{P(k)}, F(\sigma, k)\},$$

$$e(\overline{P(k)}, y) = e(\overline{P(k)}, x) - \varepsilon,$$

and

$$e(F(\sigma, k), y) = e(F(\sigma, k), x) + \varepsilon.$$

So $\theta(y)$ is lexicographically less than $\theta(x)$ if we consider only the coalitions in $\mathcal{H}$. This contradicts $x \in \mathcal{H}$-nucleolus. Similarly it can be shown that there does not exist a $k \in N$ such that

$$e(\overline{P(k)}, x) - e(F(\sigma, k), x) < 0.$$

Hence $e(S, x) = e(N \setminus S, x)$ for all $S \in \mathcal{H}$. With theorem 4.2.34 it follows that $x = EGS(\sigma; \alpha; s)$. $\qquad\square$

Using a result of Potters and Reinierse [100] which we will discuss in more detail in the next section, we can give necessary and sufficient conditions for the nucleolus of a sequencing game to equal the allocation generated by the *EGS*-rule. First we define $\bar{s}_{ij}(x)$ by

$$\bar{s}_{ij}(x) = \max\{v(T) - x(T) | i \in T \subset N \setminus \{j\}, T \text{ is connected}\}.$$

Potters and Reijnierse show that the nucleolus of a sequencing game is the unique vector x satisfying efficiency and $\bar{s}_{ii+1}(x) = \bar{s}_{i+1i}(x)$ for all $i \in N \setminus \{\sigma^{-1}(n)\}$. For notational convenience's sake we will assume, w.l.o.g. in the formulation and proof of the following theorem, that σ is equal to the identity. That is, the original order of the players is 123...n. Also for convenience's sake we introduce the notation

$$[k, p] := \{j \in N | k \leq j \leq p\}.$$

Theorem 4.2.36 *Let v be a sequencing game corresponding to the sequencing situation $(\sigma; \alpha; s)$. Then $EGS(\sigma; \alpha; s) = \nu(v)$ if and only if for all $i \in N$*

$$\sum_{j=1}^{p-1} \sum_{k=p}^{i} g_{jk} \geq \sum_{j=1}^{p-1} \sum_{k=i+1}^{n} g_{jk} \quad \text{for all } 1 < p \leq i$$
$$\sum_{h=i+1}^{q} \sum_{l=q+1}^{n} g_{hl} \geq \sum_{h=1}^{i} \sum_{l=q+1}^{n} g_{hl} \quad \text{for all } i+1 \leq q < n.$$

Proof. The inequalities stated in the theorem are equivalent to

$$\sum_{j=1}^{p-1} \sum_{k=p}^{i} g_{jk} \geq \sum_{j=1}^{i} \sum_{k=i+1}^{n} g_{jk} - \sum_{j=p}^{i} \sum_{k=i+1}^{n} g_{jk}$$
$$\sum_{h=i+1}^{q} \sum_{l=q+1}^{n} g_{hl} \geq \sum_{h=1}^{i} \sum_{l=i+1}^{n} g_{hl} - \sum_{h=1}^{i} \sum_{l=i+1}^{q} g_{hl}$$

for all $1 < p \leq i$ and all $i+1 \leq q < n$, respectively.
Multiplying by $-\frac{1}{2}$ and rearranging yield the following equivalent inequalities.

$$-\tfrac{1}{2} \sum_{j=1}^{i} \sum_{k=i+1}^{n} g_{jk} \geq -\tfrac{1}{2} \sum_{j=p}^{i} \sum_{k=i+1}^{n} g_{jk} - \tfrac{1}{2} \sum_{j=1}^{p-1} \sum_{k=p}^{i} g_{kj}$$
$$-\tfrac{1}{2} \sum_{h=1}^{i} \sum_{l=i+1}^{n} g_{hl} \geq -\tfrac{1}{2} \sum_{h=1}^{i} \sum_{l=i+1}^{q} g_{hl} - \tfrac{1}{2} \sum_{h=i+1}^{q} \sum_{l=q+1}^{n} g_{hl}$$

for all $1 < p \leq i$ and all $i+1 \leq q < n$, respectively.
From the definition of the EGS-rule it follows that these are equivalent to

$$v(\overline{P(i)}) - x(\overline{P(i)}) \geq v([p,i]) - x([p,i]) \quad \text{for all } 1 < p \leq i$$
$$v(F(\sigma,i)) - x(F(\sigma,i)) \geq v([i+1,q]) - x([i+1,q]) \quad \text{for all } i+1 \leq q < n$$

where x is the allocation generated by the EGS-rule. From the definition of $\bar{s}_{ii+1}(x)$ and $\bar{s}_{i+1i}(x)$ and the fact that the EGS-rule satisfies the fair head-tail split property it follows that this is equivalent to

$$\bar{s}_{ii+1}(x) = \bar{s}_{i+1i}(x).$$

Since this is true for all $i \in N \setminus \{n\}$ it follows from the result of Potters and Reijnierse [100] that this is equivalent to $EGS(\sigma; \alpha; s) = \nu(v)$ □

Example Let $N = \{1, 2, 3, 4\}$ and let σ be the identity permutation, let $\alpha = (3, 6, 1, 2)$, and $s = (1, 1, 1, 1)$. Then the conditions of the theorem are satisfied and $EGS(\sigma; \alpha; s) = \nu(v) = (1\frac{1}{2}, 1\frac{1}{2}, \frac{1}{2}, \frac{1}{2})$.

Remark Comparing sequencing games and permutation games we see that in both situations $v(N)$ is obtained by maximizing over the set of Π_N. In permutation games we are looking at situations where every player has the same service time and where the revenue functions need not be linear, whereas in sequencing games the players need not have the same service time and the cost functions are always linear. However, in general it is not possible to define the k_{ij}'s, as used in the definition of permutation games, for sequencing games. This is due to the fact that in a permutation game the revenue of player i depends only on player i's position since all players have the same service time, while in a sequencing game it is also important to know the service times of the players standing in front of i. Consequently, for permutation games it is possible to give a linear extension of the function that we are maximizing on the set of permutation matrices to the set of doubly stochastic matrices, while this need not be possible for a sequencing game as the following example shows.

Example Let $N = \{1, 2, 3\}$, σ is the identity permutation, $\alpha = (10, 20, 30)$, $s = (7, 3, 5)$. Consider the six permutation matrices corresponding to the six permutations of N.

$$P_1 = \begin{pmatrix} 1 & 0 & 0 \\ 0 & 1 & 0 \\ 0 & 0 & 1 \end{pmatrix} \quad P_2 = \begin{pmatrix} 1 & 0 & 0 \\ 0 & 0 & 1 \\ 0 & 1 & 0 \end{pmatrix} \quad P_3 = \begin{pmatrix} 0 & 1 & 0 \\ 1 & 0 & 0 \\ 0 & 0 & 1 \end{pmatrix}$$

$$P_4 = \begin{pmatrix} 0 & 1 & 0 \\ 0 & 0 & 1 \\ 1 & 0 & 0 \end{pmatrix} \quad P_5 = \begin{pmatrix} 0 & 0 & 1 \\ 1 & 0 & 0 \\ 0 & 1 & 0 \end{pmatrix} \quad P_6 = \begin{pmatrix} 0 & 0 & 1 \\ 0 & 1 & 0 \\ 1 & 0 & 0 \end{pmatrix}$$

Then the function k to be maximized on $\{P_1, P_2, P_3, P_4, P_5, P_6\}$ in order to find $v(N)$ has the following values: $k(P_1) = 0$, $k(P_2) = -10$, $k(P_3) = 110$, $k(P_4) = 150$, $k(P_5) = 270$, and $k(P_6) = 260$. Let D be the doubly stochastic matrix defined by $D = \frac{1}{3}P_1 + \frac{1}{3}P_4 + \frac{1}{3}P_5$ then $D = \frac{1}{3}P_2 + \frac{1}{3}P_3 + \frac{1}{3}P_6$. For k to have a linear extension on the set of doubly stochastic matrices $\frac{1}{3}k(P_1) + \frac{1}{3}k(P_4) + \frac{1}{3}k(P_5)$ has to equal $\frac{1}{3}k(P_2) + \frac{1}{3}k(P_3) + \frac{1}{3}k(P_6)$. But the first expression is equal to 140 and the last is equal to 120. So it is impossible to extend k to a linear function on the set of doubly stochastic matrices.

4.3 σ-PAIRING GAMES AND σ-COMPONENT ADDITIVE GAMES

Let us consider the following game which looks a lot like a sequencing game. $N = \{1, 2, 3, 4\}$, σ is the identity permutation, and $g_{12} = g_{13} = g_{14} = g_{23} = g_{24} = g_{34} = 1$. The game v is defined as in 4.1 and 4.2. We would be inclined to say that v is a sequencing game but some further analysis show that it is not possible to find costs $\alpha_1, \alpha_2, \alpha_3, \alpha_4$ and service times s_1, s_2, s_3, s_4 such that

$g_{ij} = 1$ for all i, j with $i < j$. So v is not a sequencing game. We call v a
σ-*pairing game*.
We extend the definition of $[k, p]$ given in section 4.2 to situations where σ is
not the identity permutation.

$$[k, p]_\sigma := \{j \in N | \sigma(k) \leq \sigma(j) \leq \sigma(p)\}$$

Definition 4.3.1 *A cooperative game v is called a σ-pairing game if v is an
element of the positive cone generated by the unanimity games $u_{[i,j]_\sigma}$ with $\sigma(i) <
\sigma(j)$.*

It is clear that every sequencing game is a σ-pairing game for some σ, while
the example discussed above shows that the converse statement is not true. All
the results derived for sequencing games in the previous section also hold for
σ-pairing games.
A further generalization is given by the following class of games.

Definition 4.3.2 *Let $\sigma \in \Pi_N$. A cooperative game v is called a σ-component
additive game if*

(a) $v(i) = 0$ *for all $i \in N$,*

(b) v *is superadditive,*

(c) $v(S) = \sum_{T \in S/\sigma} v(T)$.

σ-pairing games and σ-component additive games were introduced in [24].
Straightforward verification shows that every σ-pairing game is a σ-component
additive game, but the converse statement is not true as the following example
shows.

Example Mrs. Leroy, Mr. Mason, and Mrs. Nelson each one needs to have a
job completed on the same machine. Mrs. Leroy was the first to hand in the
job, Mr. Mason was the second and Mrs. Nelson was the last. Mrs. Leroy's
job will take one hour, Mr. Mason's job will take 3 hours, and Mrs. Nelson's
job will take 2 hours. Mrs. Leroy's job is due after 5 hours, if she is late she
will have to pay a penalty of $100. Mr. Mason's job is due after 3 hours, if he
is late he will have to pay a penalty of $200. Mrs. Nelson's job is due after 5
hours, if she is late she will have to pay a penalty of $200. If the jobs are done

S	$v(S)$	S	$v(S)$	S	$v(S)$	S	$v(S)$
$\{1\}$	0	$\{3\}$	0	$\{1,3\}$	0	$\{1,2,3\}$	300
$\{2\}$	0	$\{1,2\}$	200	$\{2,3\}$	200	$\emptyset$	0

Table 4.3 A σ-component additive game.

in the order in which they were handed in both Mr. Mason and Mrs. Nelson will be late, resulting in a total penalty of \$400. They can achieve a maximal cost savings of \$300 by doing first Mr. Mason's job, then Mrs. Nelson's job, and finally, Mrs. Leroy's job. We can consider the maximal cost savings that each coalition can achieve by shifting its members without jumping over non-members. The game v that we obtain in this way is given in table 4.3. Since this game is not convex it is not a σ-pairing game. However, it is a σ-component additive game.

For σ-component additive games we define a division rule, the β-rule, which is an extension of the *EGS*-rule.

Definition 4.3.3 *Let v be a σ-component additive game. The β-rule is given by*

$$\beta_i(v) := \frac{1}{2}(v(\overline{P(i)}) - v(P(\sigma, i)) + v(F(\sigma, i) \cup \{i\}) - v(F(\sigma, i))).$$

From lemma 4.2.31 it follows that the β-rule is indeed an extension of the *EGS*-rule. The following theorem shows that, just like the *EGS*-rule does for σ-pairing games, the β-rule generates a core element for σ-component additive games.

Theorem 4.3.4 *Let v be a σ-component additive game. Then $\beta(v) \in C(v)$.*

Proof. Straightforward verification shows that $\sum_{i \in N} \beta_i(v) = v(N)$. Let T be a connected coalition, say $T = [i, j]_\sigma$. Then

$$\sum_{k \in T} \beta_k(v) = \frac{1}{2}[v(F(\sigma, i) \cup \{i\}) - v(P(\sigma, i))]$$

$$+ \frac{1}{2}[v(\overline{P(j)}) - v(F(\sigma, j))]$$

$$= \frac{1}{2}[v(T \cup [j+1, n]_\sigma) - v([j+1, n]_\sigma)]$$

$$+ \quad \frac{1}{2}[v(T \cup [1, i-1]_\sigma) - v([1, i-1]_\sigma)]$$

$$\geq \quad \frac{1}{2}[v(T) + v(T)] = v(T)$$

where the inequality follows from the superadditivity of v. Because of (c) in the definition of a σ-component additive game, it follows that $\sum_{i \in N} \beta_i(v) \geq v(S)$ for every coalition S. $\qquad\square$

Potters and Reinierse [100] studied the following generalization of σ-component additive games which they called Γ-*component additive games*. They consider a tree Γ, i.e., a connected graph without cycles. The nodes of the graphs represent the players. A coalition is connected if the set of nodes representing T in Γ is connected, that is, if for any two nodes $i, j \in T$ there is a path in Γ from i to j which uses only nodes in T. A component of a non-connected coalition is a maximal (with respect to set inclusion) connected subset of S. The set of components of S in Γ is denoted by S/Γ. If there is an edge connecting i and j in Γ we say that $(i, j) \in \Gamma$. A Γ-component additive game is defined in the same way as a σ-component additive game with S/Γ replacing S/σ. It is clear that every σ-component additive game is a Γ-component additive game by taking Γ to be the line graph $\sigma^{-1}(1) - \sigma^{-1}(2) - \ldots - \sigma^{-1}(n)$.
Potters and Reijnierse show that Γ-component additive games are balanced. In the following we will discuss the result that they obtain with respect to the bargaining set, kernel, and nucleolus of Γ-component additive games. To proof their result concerning the bargaining set we will need the following lemma.

Lemma 4.3.5 (Potters and Reijnierse) *Let v be a cooperative game with $I(v) \neq \emptyset$. Let $x \in I(v) \setminus C(v)$. Then there exist a coalition S_0 and a vector $z \in R_+^n$ such that:*

(a) $z_j = 0$ *for $j \notin S_0$,*

(b) $z(T) \geq e(T, x)$ *for all $T \in 2^N \setminus \{\emptyset\}$,*

(c) $z(S_0) = e(S_0, x)$,

(d) $e(T, x) < 0$ *for all coalitions T with $T \cap S_0 = \emptyset$.*

Proof. First the vector z and a coalition T_0 which contains all the players i with $z_i > 0$ is constructed. This construction is broken down in two steps.
The construction of z

Start with an arbitrary node r and $T_0 = \emptyset$. The choice of r induces a partial ordering on Γ as follows: $j \succeq i$ if i is on the unique path from r to j in Γ. We define $\Gamma_{\succeq i} := \{j \in N | j \succeq i\}$. We start with defining $z_k = 0$ for all nodes k with degree 1 in Γ. Also, k is added to T_0 if and only if $e(\{k\}, x) = 0$. As soon as all the z_j with $j \succ i$ have been defined we define z_i by $z_i := \max\{0, \bar{z}_i\}$ where

$$\bar{z}_i := \max\{e(S, x) - z(S \setminus \{i\}) | S \text{ connected}, i \in S \subset \Gamma_{\succeq i}\}.$$

Player i is added to T_0 if and only if $\bar{z}_i \geq 0$. For each $i \in T_0$ there is a connected coalition S_i with $i \in S_i \subset \Gamma_{\succeq i}$ and $z(S_i) = e(S_i, x)$. The vector $z \geq 0$ will be completely defined after finitely many steps. It satisfies $z(S) \geq e(S, x)$ for all connected S and therefore also for all non-connected non-empty coalitions. Also, for every $i \in T_0$ it satisfies $z(S_i) = e(S_i, x)$.

The construction of S_0

We have $T_0 \subset \bigcup_{i \in T_0} S_i$. We start pruning the collection $\{S_i\}_{i \in T_0}$ to get rid of redundancy. We scan the coalitions S_i to see which one we can delete. We start with the coalitions S_i with the property that the path from r to i does not contain a point other than i in T_0. These coalitions are labeled, $S_i \to S_i^0$. If, for example, r is in T_0, only the coalition S_r will initially obtain a label. If $j \in S_i^0$ for some $i \prec j$ we delete S_j. If all nodes $i \in T_0$ with $i \prec j$ have been scanned and S_j has not been deleted we label coalition S_j. Proceeding in this way all coalitions S_i with $i \in T_0$ will have been deleted or labeled. Let T_1 be the subset of T_0 which contains all i for which S_i has been labeled. The collection $\{S_i\}_{i \in T_1}$ still covers T_0. Let $S_i, S_j \in \{S_i\}_{i \in T_1}$. Suppose $k \in S_i \cap S_j$. The path from r to k contains both i and j. W.l.o.g. we assume that $i \prec j$. Then the path from i to k contains j and since S_i is connected and $i, k \in S_i$ it follows that $j \in S_i$. But this contradicts $j \in T_1$. So $S_i \cap S_j = \emptyset$ for all $i, j \in T_0$. Define $S_0 = \bigcup_{i \in T_1} S_i$. We will show that S_0 and z satisfy properties (a)–(d).

(a) By the way that we defined S_0 and z we have $z_j = 0$ for $j \notin S_0$.

(b) By the definition of z.

(c) $z(S_0) = \sum_{i \in T_1} z(S_i) = \sum_{i \in T_1} e(S_i, x) \leq e(S_0, x) \leq z(S_0)$.

(d) Let T be a connected coalition with $T \cap S_0 = \emptyset$. Then $z(T) = 0 \geq e(T, x)$. Suppose $e(T, x) = 0$. Let $i \in T$ be such that there is no other node in T on the path form r to i. Then $i \in T_0 \subset S_0$ because of the way z and T_0 have been constructed. This contradicts the fact that T and S_0 are disjoint. So $e(T, x) < 0$. $\qquad\square$

Theorem 4.3.6 (Potters and Reijnierse) *Let v be a Γ-component additive game. Then $\mathcal{M}(v) = C(v)$.*

Proof. Let $x \in I(v) \setminus C(v)$. Let S_0 and $z \in R_+^n$ be as in lemma 4.3.5. Since $x \notin C(x)$ there is an $S \subset N$ with $z(S) \geq e(S, x) > 0$. So $z \neq 0$. Also,

$e(S_0, x) = z(S_0) = z(N) > 0$, together with $e(N, x) = 0$ this implies that $S_0 \neq N$. Therefore, we can consider i, j with $i \in S_0$ and $z_i > 0$, and $j \notin S_0$. We can define an objection $(y; S_0,)$ of player i against player j as follows. Define

$$y_k = x_k + z_k + \frac{z_i}{|S_0|} \quad \text{for } k \in S_0 \setminus \{i\}$$
$$y_i = x_i + \frac{z_i}{|S_0|}$$

Then $y(S_0) = x(S_0) + z(S_0) = v(S_0)$ and $y_l > x_l$ for all $l \in S_0$. So $(y; S_0)$ is an objection of i against j. Let $T \subset N \setminus \{i\}$. If $T \cap S_0 = \emptyset$ then $e(T, x) < 0$ and therefore T cannot be used in any counter objection. If $T \cap S_0 \neq \emptyset$ then $z(T \cap S_0) = z(T) \geq e(T, x)$ and it follows that

$$
\begin{aligned}
y(T \cap S_0) + x(T \setminus S_0) &> x(T \cap S_0) + z(T \cap S_0) + x(T \setminus S_0) \\
&= x(T) + z(T) \geq v(T)
\end{aligned}
$$

So there is no counter objection to $(y; S_0)$ and it follows that $x \notin \mathcal{M}(v)$ and $\mathcal{M}(v) \subset C(v)$. Since the core is always a subset of the bargaining set it follows that $\mathcal{M}(v) = C(v)$. □

For the result on the kernel of a Γ-component additive game we need the following theorem which we will state without a proof.

Theorem 4.3.7 (Kohlberg) *Let v be a zero-normalized game with $I(v) \neq \emptyset$. Let $x \in I(v)$. Then $x = \nu(v)$ if and only if for every $t \in R$ the collection $\mathcal{B}_t = \{S \subset N, S \neq \emptyset, N | e(S, x) \geq t\}$ is balanced on the set $\{i \in N | x_i > 0\}$ or empty.*

Theorem 4.3.8 (Potters and Reijnierse) *Let v be a Γ-component additive game. Then $\mathcal{K}(v) = \{\nu(v)\}$.*

Let $x \in \mathcal{K}(v)$, then $x \in C(v)$. Let $t \in R$ and suppose $\mathcal{B}_t \neq \emptyset$. $S \in \mathcal{B}_t$ implies $T \in \mathcal{B}_t$ for all $T \in S/\Gamma$ since all excesses are negative. If $T_1 \in \mathcal{B}_t$ is connected and $i \in T_1 \subset N \setminus \{j\}$ with $(i, j) \in \Gamma$ there is a connected $T_2 \in \mathcal{B}_t$ with $j \in T_2 \subset N \setminus \{i\}$. (Since $s_{ij}(x) = s_{ji}(x)$.) Note that $T_1 \cap T_2 = \emptyset$. We can continue this process with $T_1 \cup T_2$. In this way we construct a partition $\{T_1, T_2, \ldots, T_k\}$ of N with $T_l \in \mathcal{B}_t$ for all $1 \leq l \leq k$. This implies that $\mathcal{B}_t$ is the union of balanced collections and is therefore balanced itself. So by theorem 4.3.7 $x = \nu(v)$. □

Now we can proof the result used in theorem 4.2.36.

Theorem 4.3.9 *Let v be Γ-component additive game. Let $x \in R^n$. Then*

$$x = \nu(v) \text{ if and only if } x \in PI(v) \text{ and } \bar{s}_{ij}(x) = \bar{s}_{ji}(x) \text{ for all } (i,j) \in \Gamma.$$

Proof. If $x = \nu(v)$ then the left hand statement holds since $x \in \mathcal{PK}(v)$.
Suppose $x \in PI(v)$ with $\bar{s}_{ij}(x) = \bar{s}_{ji}(x)$ for all $(i,j) \in \Gamma$. If $x \in C(v)$ then
$x = \nu(v)$ since $\mathcal{PK}(v) = \mathcal{K}(v) = \{\nu(v)\}$. Suppose $x \notin C(v)$. Let $\mathcal{B}$ be the
collection containing all connected coalitions with positive excess. Take $S \in \mathcal{B}$
with maximal excess.
If $(i,j) \in \Gamma$ and $i \in S \subset N \setminus \{j\}$, then $\bar{s}_{ij}(x) > 0$ and hence $\bar{s}_{ji}(x) > 0$. So
there is a coalition $T \in \mathcal{B}$ with $j \in T \subset N \setminus \{i\}$. Then $S \cup T$ is connected and
$S \cap T = \emptyset$. But $e(S \cup T, x) > e(S, x)$, which contradicts the choice of S. So
there is no $x \in PI(v) \setminus C(v)$ that satisfies $\bar{s}_{ij}(x) = \bar{s}_{ji}(x)$ for all $(i,j) \in \Gamma$. $\square$

5

TRAVELLING SALESMAN GAMES
AND ROUTING GAMES

5.1 INTRODUCTION

Four universities, one in Ohio, one in Pennsylvania, one in Quebec, and one in Rhode Island are considering to invite a speaker from Paris, France, to give lectures at each of the universities. They are studying how to make the costs of travel as low as possible. It is clear to all of them that it does not make sense for each one of them to pay a two way ticket from Paris and back. Once the speaker is in North America they should let him visit each university and then travel back. They want to find an order for him to visit the universities that minimizes the total travel costs. The costs for a one way ticket between Paris and the universities in either direction are: for Ohio, $350, for Pennsylvania, $300, for Quebec, $200, and for Rhode Island, $250. The costs for a one way ticket between the universities in either directions are: Ohio-Pennsylvania, $100, Ohio-Quebec, $150, Ohio-Rhode Island, $150, Pennsylvania-Quebec, $150, Pennsylvania-Rhode Island, $100, and Quebec-Rhode Island, $100. These costs are summarized in table 5.1. After some

	Paris	Ohio	Pennsylvania	Quebec	Rhode Island
Paris	0	350	300	200	250
Ohio	350	0	100	150	150
Pennsylvania	300	100	0	150	100
Quebec	200	150	150	0	100
Rhode Island	250	150	100	100	0

Table 5.1 The costs of travelling between Paris and the universities, and between the universities.

S	$c(S)$	S	$c(S)$	S	$c(S)$	S	$c(S)$
{1}	70	{1,2}	75	{2,4}	65	{1,3,4}	75
{2}	60	{1,3}	70	{3,4}	55	{2,3,4}	70
{3}	40	{1,4}	75	{1,2,3}	75	{1,2,3,4}	85
{4}	50	{2,3}	65	{1,2,4}	80	$\emptyset$	0

Table 5.2 The travelling salesman game arising from the example of the introduction.

analysis the universities realize that they will minimize the travel costs if the let the speaker travel form Paris to Pennsylvania, to Ohio, to Rhode Island, to Quebec, and then back to Paris. With this tour the total travel costs are $850. Now they still need a way to divide these costs among them. Simply assigning the cost pertaining to each leg of the tour to the university at which that leg ends will not work, because the speaker will have to return to Paris and the universities will have to cover those costs too. Furthermore, it seems to be putting an undue burden on the first university to be visited. They decide to perform a game theoretic analysis of this problem. Each coalition of universities figures out in what order the speaker should visit the members of the coalition so as to minimize travel costs. The ensuing game is given in table 5.2. There Ohio=1, Pennsylvania=2, Quebec=3, and Rhode Island=4. This is a cost game, so the reader should keep in mind that each university would like to contribute as little as possible to the total costs. This game has a non-empty core. A core element is, for example, $(27\frac{1}{2}, 22\frac{1}{2}, 12\frac{1}{2}, 22\frac{1}{2})$. This cost allocation is obtained by dividing the travel costs from and to Paris equally among the four universities, and making university i pay all of the travel costs from university i to university j. The allocation that we obtain if university j pays all the travel costs from university i to university j and the travel costs from and to Paris are divided equally among the four universities, that is, the allocation $(22\frac{1}{2}, 12\frac{1}{2}, 27\frac{1}{2}, 22\frac{1}{2})$ is also an element of the core of the game. This game is an example of a *travelling salesman game* which we will study in more detail in this chapter. In section 5.2 we will introduce travelling salesman games formally. We will show that, in general, a travelling salesman game need not be balanced, and we will consider special classes of travelling salesman games that contain only balanced games. In section 5.3 we will introduce a class of games that are related to travelling salesman games, the routing games. Routing games are balanced under very weak conditions. We will study a class of cost or weight matrices for which that generates travelling salesman games that are equal to routing games.

5.2 TRAVELLING SALESMAN GAMES

The *travelling salesman problem* is a well known combinatorial optimization problem. Except for situations similar to the one discussed in the introduction, several problems that at first look quite different can be modelled as travelling salesman problems, cf. [79]. The travelling salesman problem can be formulated as follows: Given a directed graph with weights on the arcs, find a cycle which visits each node exactly once and has minimal weight. The weight of a cycle is the sum of the weights of its arcs. In an arbitrarily given graph there need not exist a cycle which visits each vertex exactly once. Even when such a cycle exists there may be a cycle with lesser weight which visits some vertices more than once. However, if the graph is complete and the weights satisfy the *triangle inequality*, then there is a cycle in the graph, which visits each vertex exactly once, and such that no cycle which visits a vertex more than once, has lesser weight.

Definition 5.2.1 *Let G be a directed graph with w_{ij} denoting the weight of the arc starting in i and ending in j. The weights are said to satisfy the triangle inequality if*

$$w_{ij} \leq w_{ik} + w_{jk} \text{ for all vertices } i, j, k.$$

Every travelling salesman problem on a connected graph can be transformed into a problem on a complete graph with weights satisfying the triangle inequality. One defines the complete graph on the same set of vertices and one takes the weight w_{ij} in the new graph to be the weight of a minimal path from i to j in the original graph. In the following we will assume that the travelling salesman problems we discuss are on complete graphs with weights satisfying the triangle inequality, unless explicitly stated otherwise.

By associating all but one of the vertices with players we obtain a cooperative cost game. Let us denote the vertex that does not correspond to a player by 0. In this game each coalition S wants to construct a tour which starts in 0, visits each member of S and exactly once, and returns to 0, and which has minimal weight. In other words, S wants to find a minimal weight travelling salesman tour on the complete graph with set of vertices equal to $S \cup \{0\}$. We can use a bijection $e : \{1, \ldots, |S|\} \to S$ to describe a tour on $S \cup \{0\}$. The tour corresponding to such a bijection e starts in 0, then visits $e(1)$, then $e(2)$, etc. The last vertex in S that it visits is $e(|S|)$ after which it returns to 0. The weight of such a tour is

$$w_{0e(1)} + w_{e(1)e(2)} + \cdots + w_{e(|S|-1)e(|S|)} + w_{e(|S|)0}.$$

	Idle	Solomon	Temploy	Usprint	Vandam	p_i
Idle	0	1	2	2	1	
Solomon	1	0	1	2	2	0.5
Temploy	2	1	0	1	2	0.3
Usprint	1	2	2	0	2	0.4
Vandam	1	2	1	1	0	0.2

Table 5.3   The transition and processing costs for the multi-purpose machine.

Let $E(S)$ denote the set of all bijections from $\{1, \ldots, |S|\}$ to S. The cost $c(S)$ of a coalition S is given by

$$c(S) := \min_{e \in E(S)} \left(w_{0e(1)} + w_{e(1)e(2)} + \cdots + w_{e(|S|)0} \right). \tag{5.1}$$

The game defined by 5.1 is called a *travelling salesman game*.
Let $S_0 := S \cup \{0\}$. Another expression for $c(S)$ in a travelling salesman game is given by the following mixed programming problem. (Cf. Tucker [137])

$$
\begin{aligned}
\min \quad & \sum_{i \in S_0} \sum_{j \in S_0} w_{ij} x_{ij} \\
\text{s.t.} \quad & \sum_{j \in S_0 \setminus \{i\}} x_{ij} = 1_{S_0}(i) && \text{for all } i \in N_0 \\
& \sum_{i \in S_0 \setminus \{j\}} x_{ij} = 1_{S_0}(j) && \text{for all } j \in N_0 \\
& u_i - u_j + n x_{ij} \leq n - 1 && \text{for all } i, j \in S \text{ with } i \neq j \\
& x_{ij} \in \{0, 1\}, \; u_i \in R && \text{for all } i, j \in N_0.
\end{aligned}
\tag{5.2}
$$

In terms of the problem of the universities and the speaker $x_{ij} = 1$ means that the speaker visits university j immediately after university i. The first $2n + 2$ constraints guarantee that the speaker leaves his home town 0 and visits each university exactly once and returns to his home town, while the inequalities guarantee that the travel plan described by the x_{ij}'s does not contain subtours. The following example discusses a travelling salesman game that arises from what seems to be a completely different situation.

Example Four companies, Solomon, Temploy, Usprint, and Vandam are using a multi-purpose machine to perform different tasks for each of them. Solomon's task is denoted by t_1, Temploy's by t_2, Usprint's by t_3, and Vandam's by t_4. The machine can be in several states. When it is idle it should be in state s_0. To be able to perform task t_i the machine has to be brought to state s_i. The machine can be brought from any state into any other state. The transition costs are denoted by w_{ij} for $i, j \in \{0, 1, 2, 3, 4\}$. The processing costs of performing task t_i are denoted by p_i. These transition and processing costs (in \$1000) are given in table 5.3. If a company does not cooperate with any other company, it alone

S	$c(S)$	S	$c(S)$	S	$c(S)$	S	$c(S)$
{1}	2.5	{1,2}	4.8	{2,4}	4.5	{1,3,4}	6.1
{2}	4.3	{1,3}	4.9	{3,4}	3.6	{2,3,4}	4.9
{3}	3.4	{1,4}	4.7	{1,2,3}	5.2	{1,2,3,4}	7.4
{4}	2.2	{2,3}	4.7	{1,2,4}	5.0	∅	0

Table 5.4 The game arising from the multi-purpose machine problem.

has to cover, the transition costs of bringing the machine from the idle state to the necessary state to perform its task and back to the idle state, plus the processing cost. By cooperating the companies can reduce the costs incurred. There is nothing they can do about the processing costs but they can try to find an order of doing the tasks that will minimize the transition costs. A game theoretic analysis of this problem reveals that the cost $c(S)$ of coalition S is the sum of $p(S)$ and the minimum in 5.1 or 5.2. The game c is given in table 5.4. Here Solomon=1, Temploy=2, Usprint=3, and Vandam=4. The game is the sum of a travelling salesman game $\hat{c}$ and the additive game p. The core of c is empty. Take the balanced collection $\{\{1,2,3\},\{1,2,4\},\{3,4\}\}$ with all weights equal to $\frac{1}{2}$. Then

$$\frac{1}{2}(c(1,2,3) + c(1,2,4) + c(3,4)) = 6.9 < 7.4 = c(1,2,3,4).$$

So c is not balanced. Therefore, the travelling salesman game $\hat{c}$ also has an empty core.

The above example shows that a travelling salesman game need not be balanced. However, if there are three or less players, the travelling salesman game will have a non-empty core.

Proposition 5.2.2 *Let c be a travelling salesman game with three or less players, then c is balanced.*

Proof. If c is a game with one player, c is trivially balanced. If c has two players then c is balanced because c is subadditive. Let c be a travelling salesman game with three players. To show that c is balanced we only have to show that

$$c(1,2) + c(1,3) + c(2,3) \geq 2c(1,2,3). \tag{5.3}$$

We will distinguish two cases.
(i) In the expression $c(1,2) + c(1,3) + c(2,3)$ there is a w_{0i} occurring twice.

W.l.o.g. we assume that w_{01} occurs twice. Then

$$
\begin{aligned}
c(1,2) + c(1,3) + c(2,3) &= w_{01} + w_{12} + w_{20} + w_{01} + w_{13} + w_{30} \\
&+ \min\{w_{02} + w_{23} + w_{30}, w_{03} + w_{32} + w_{20}\}.
\end{aligned}
$$

Suppose the minimum is equal to $w_{02} + w_{23} + w_{30}$. The other case can be treated similarly. Then

$$
\begin{aligned}
c(1,2) + c(1,3) + c(2,3) &= w_{01} + w_{12} + w_{20} + w_{01} + w_{13} \\
&+ w_{30} + w_{02} + w_{23} + w_{30} \\
&= w_{01} + w_{12} + w_{23} + w_{30} + w_{02} \\
&+ w_{20} + w_{01} + w_{13} + w_{30} \\
&\geq c(1,2,3) + w_{02} + w_{20} + w_{01} + w_{13} + w_{30} \\
&\geq c(1,2,3) + w_{02} + w_{21} + w_{13} + w_{30} \\
&\geq 2c(1,2,3).
\end{aligned}
$$

(ii) In the expression $c(1,2) + c(1,3) + c(2,3)$ there is no w_{0i} occurring twice. Then

$$
c(1,2) + c(1,3) + c(2,3) = w_{01} + w_{12} + w_{20} + w_{03} + w_{31} + w_{10} + w_{02} + w_{23} + w_{30}
$$

or

$$
c(1,2) + c(1,3) + c(2,3) = w_{02} + w_{21} + w_{10} + w_{01} + w_{13} + w_{30} + w_{03} + w_{32} + w_{20}.
$$

Both cases can be treated similarly. Consider the first case. There

$$
\begin{aligned}
c(1,2) + c(1,3) + c(2,3) &= w_{01} + w_{12} + w_{23} + w_{30} + w_{20} \\
&+ w_{02} + w_{03} + w_{31} + w_{10} \\
&\geq c(1,2,3) + w_{20} + w_{02} + w_{03} + w_{31} + w_{10} \\
&\geq c(1,2,3) + w_{02} + w_{23} + w_{31} + w_{10} \\
&\geq 2c(1,2,3).
\end{aligned}
$$

We see that 5.3 holds in all cases and hence c is balanced.   □

The travelling salesman game $\hat{c}$ from the example of table 5.4 does not arise from a symmetric travelling salesman problem. A travelling salesman problem is called *symmetric* if $w_{ij} = w_{ji}$ for all $i, j \in N_0$. For a travelling salesman game arising from a symmetric problem Tamir [129] showed that if it has 4 or less players it will have a non-empty core, and Kuipers [76] showed that also with

5 players the core will always be non-empty. Tamir gave an example of a non-balanced travelling salesman game with six players, arising from a symmetric problem.

In the following we will discuss some classes of travelling salesman games with non-empty cores. It seems intuitively clear that when the home town of the speaker is relatively far away from the universities, while these are clustered together, causing the travel costs of a trip visiting all universities to be rather cheap compared with the costs of travel from and to the home town, it is so much more profitable for the universities to cooperate, that they should be able to distribute the travel costs in such a way that no coalition will have an incentive to separate and work on its own. The following theorem formalizes this.

Theorem 5.2.3 *Let c be a travelling salesman game such that for all $i, j \in N$*

$$\frac{w_{0i} + w_{j0}}{n} \geq \max_{e \in E(N)} \left(w_{e(1)e(2)} + w_{e(2)e(3)} + \cdots + w_{e(n-1)e(n)}\right).$$

Then c is balanced.

Proof. W.l.o.g. we assume that

$$c(N) = w_{01} + w_{12} + \cdots + w_{n0}.$$

Let $x \in R^n$ be defined by

$$x_i = w_{ii+1} + (w_{01} + w_{n0})/n \quad \text{for } i \in N \setminus \{n\}$$
$$x_n = (w_{01} + w_{n0})/n.$$

Then $x(N) = c(N)$. Let $S \subset N$, $S \neq N$. Let $e \in E(S)$ be such that

$$c(S) = w_{0e(1)} + w_{e(1)e(2)} + \cdots + w_{e(|S|-1)e(|S|)} + w_{e(|S|)0}.$$

In the following we will use the notations

$$M_S := w_{e(1)e(2)} + w_{e(2)e(3)} + \cdots + w_{e(|S|-1)e(|S|)} \qquad \text{and}$$
$$M := \max_{f \in E(N)}\left(w_{f(1)f(2)} + w_{f(2)f(3)} + \cdots + w_{f(n-1)f(n)}\right).$$

Then

$$\begin{aligned}
c(S) &= w_{0e(1)} + w_{e(|S|)0} + M_S \\
&= \frac{|S|}{n}\left(w_{0e(1)} + w_{e(|S|)0}\right) + \frac{n - |S|}{n}\left(w_{0e(1)} + w_{e(|S|)0}\right) + M_S
\end{aligned}$$

S	$c(S)$	S	$c(S)$	S	$c(S)$	S	$c(S)$
{1}	25.5	{1,2}	25.8	{2,4}	26.5	{1,3,4}	30.1
{2}	25.3	{1,3}	28.9	{3,4}	28.6	{2,3,4}	28.9
{3}	28.4	{1,4}	28.7	{1,2,3}	29.2	{1,2,3,4}	30.4
{4}	29.2	{2,3}	28.7	{1,2,4}	28.0	$\emptyset$	0

Table 5.5 A balanced modification of the game of table 5.4.

$$
\geq \frac{S}{n}(w_{0e(1)} + w_{e(|S|)0}) + (n - |S|)M + M_S
$$

$$
= \frac{|S|}{n}(w_{0e(1)} + w_{e(|S|)0}) + \frac{|S|}{n}M + (n - |S| - \frac{|S|}{n})M + M_S
$$

$$
\geq \frac{|S|}{n}(w_{01} + w_{n0} + \sum_{i \in S\setminus\{n\}} w_{ii+1}) + (n - |S| - \frac{|S|}{n})M + M_S
$$

$$
\geq \frac{|S|}{n}(w_{01} + w_{n0} + \sum_{i \in S\setminus\{n\}} w_{ii+1}) + \frac{n - |S|}{n}M + M_S
$$

$$
\geq \frac{|S|}{n}(w_{01} + w_{n0}) + \sum_{i \in S\setminus\{n\}} w_{ii+1}
$$

$$
= x(S).
$$

Here the second inequality follows form the fact that

$$
w_{0e(1)} + w_{e(|S|)0} + M < w_{01} + w_{n0} + \sum_{i \in S\setminus\{n\}} w_{ii+1} < w_{01} + w_{n0} + \sum_{i \in N\setminus\{n\}} w_{ii+1}
$$

(5.4)

contradicts $c(N) = w_{01} + w_{12} + \cdots + w_{n0}$, since 5.4 implies that any travel plan starting with $0 - e(1)$ and ending with $e(|S|) - 0$ is cheaper. So $x \in C(c)$. $\square$

Example In the non-balanced game of table 5.4 we can make the following changes. Change w_{01} from 1 to 14, w_{02} from 2 to 15, w_{03} from 2 to 16, w_{04} from 1 to 15, w_{10} from 1 to 11, w_{20} from 2 to 10, w_{30} from 1 to 12, w_{40} from 1 to 14. The new game is given in table 5.5. This game is balanced. If we use the optimal order $0 - 1 - 3 - 4 - 2 - 0$ the core element described in the proof of theorem 5.2.3 is $(8.5, 7.3, 8.4, 6.2)$. If we use the optimal order $0 - 4 - 3 - 1 - 2 - 0$ it is $(7.75, 6.55, 8.65, 7.45)$.

The following theorem discusses another class of balanced travelling salesman games.

Theorem 5.2.4 *Let c be a travelling salesman game with the property that for each $i \in N_0$ there exist a_i, b_i such that $w_{ij} = a_i + b_j$ for all $i, j \in N_0$, with $i \neq j$. Then*

$$C(c) = \{x \in R^n | x_i = a_i + b_i + \lambda_i(a_0 + b_0) \text{ with } \lambda \in [0,1]^n \text{ and } \lambda(N) = 1\}.$$

Proof. For such a travelling salesman game all travel plans for a certain coalition S have the same cost, $\sum_{i \in S_0}(a_i + b_i)$. Let $x \in C(c)$. Then

$$x_i \leq a_i + b_i + a_0 + b_0 \qquad \text{and}$$
$$x_i \geq c(N) - c(N \setminus \{i\}) = a_i + b_i \quad \text{for all } i \in N.$$

This, and the efficiency of x imply that x is an element of the set on the right hand side. Let x be an element of this set. Then $x(N) = c(N)$ and for all $S \in 2^N$,

$$x(S) \leq \sum_{i \in S_0}(a_i + b_i) = c(S).$$

So $x \in C(c)$. $\qquad\qquad\qquad\square$

Remark An example of a travelling salesman game with the property stated in theorem 5.2.4 arises in the multi-purpose machine case when the costs incurred in changing from state i to state j can be decomposed into a cost a_i for leaving i, and a cost b_j for entering j. One can also think of a situation in which taxes are levied when entering and leaving a town. If there are no other travel costs then we obtain a game as in theorem 5.2.4.

Proposition 5.2.5 *Let c be a travelling salesman game with $w_{00} = 0$. Then there exist a_i, b_i for all $i \in N_0$ such that $w_{ij} = a_i + b_j$ for all $i, j \in N_0$ with $i \neq j$, if and only if for all $j \in N$, $w_{ij} - w_{i0}$ is the same for all $i \in N_0$.*

Proof. Let $w_{ij} = a_i + b_j$ for all $i, j \in N_0$ with $i \neq j$. Then

$$w_{ij} - w_{i0} = a_i + b_j - a_i - b_0 = b_j - b_0 \quad \text{for all } i \in N_0.$$

This proofs the "only if" part. Suppose $w_{ij} - w_{i0}$ is the same for all $i \in N_0$, say $w_{ij} - w_{i0} = b_j$. Define $a_i = w_{i0}$ for all $i \in N_0$ and $b_0 = 0$. Then $w_{ij} = a_i + b_j$

S	$\bar{c}(S)$	S	$\bar{c}(S)$	S	$\bar{c}(S)$	S	$\bar{c}(S)$
$\{1\}$	3.5	$\{1,2\}$	5.8	$\{2,4\}$	5.5	$\{1,3,4\}$	7.1
$\{2\}$	5.3	$\{1,3\}$	5.9	$\{3,4\}$	4.6	$\{2,3,4\}$	5.9
$\{3\}$	4.4	$\{1,4\}$	5.7	$\{1,2,3\}$	6.2	$\{1,2,3,4\}$	8.4
$\{4\}$	3.2	$\{2,3\}$	5.7	$\{1,2,4\}$	6.0	$\emptyset$	0

Table 5.6   A perturbation of the multi-purpose machine game.

for all $i, j \in N$. Thus, the "if" part is proved. $\square$

We can add entrance and exit taxes to the travel costs in a travelling salesman game, perturbing the game in this way. Doing this, it is possible to turn a non-balanced travelling salesman game into a balanced travelling salesman game.

Example Consider the travelling salesman game of table 5.4 with weights given in table 5.3. We perturb the game a little by introducing extra costs involved in leaving and entering state 0. The tax for leaving state 0 is $a_0 = \frac{1}{2}$, and the tax for entering state 0 is $b_0 = \frac{1}{2}$. Let $\bar{w}_{ij}$ denote the new costs of going from i to j, then

$$\begin{aligned} \bar{w}_{i0} &= w_{i0} + \tfrac{1}{2} \quad \text{for all } i \in \{1,2,3,4\} \\ \bar{w}_{0i} &= w_{0i} + \tfrac{1}{2} \quad \text{for all } i \in \{1,2,3,4\} \\ \bar{w}_{ij} &= w_{ij} \qquad\quad \text{in all other cases.} \end{aligned}$$

Let $\bar{c}$ denote the new game. This game is given in table 5.6. This game has a non-empty core. A core element is $(2.5, 1.3, 2.4, 2.2)$. Subtracting the additive game p gives us the core element $(2, 1, 2, 2)$ of the remaining travelling salesman game.

The following theorem shows that we can always perturb a travelling salesman game by adding taxes so as to make it balanced. In fact, it is sufficient to add taxes only for entering and leaving state 0. For each $a, b \in R^{N_0}$ let us define

$$u(a,b)_{ij} = \begin{cases} a_i + b_j & \text{for } i, j \in N_0,\ i \neq j \\ 0 & \text{if } i = j. \end{cases}$$

For a travelling salesman game c with weights w_{ij} let $\bar{c}$ denote the game with weights $w_{ij} + u(a,b)_{ij}$.

Theorem 5.2.6 *Let c be a travelling salesman game with weights w_{ij} for $i, j \in \{0, 1, \ldots, n\}$. Then there exist a $q(w)$ which depends only on the weights w_{ij} such that the game $\bar{c}$ is balanced if and only if $a_0 + b_0 \geq q(w)$.*

Proof. The game $\bar{c}$ is balanced if and only if

$$\sum_{S \in \mathcal{B}} \lambda_S \bar{c}(S) - \bar{c}(N) \geq 0$$

for all minimal balanced collections $\mathcal{B}$. Let $\mathcal{B}$ be a minimal balanced collection. Then

$$
\begin{aligned}
\sum_{S \in \mathcal{B}} \lambda_S \bar{c}(S) - \bar{c}(N) &= \sum_{S \in \mathcal{B}} \lambda_S (c(S) + a(S_0) + b(S_0)) \\
&\quad - (c(N) + a(N_0) + b(N_0)) \\
&= \sum_{S \in \mathcal{B}} \lambda_S (c(S) + a_0 + b_0) + a(N) + b(N) \\
&\quad - (c(N) + a(N_0) + b(N_0)) \\
&= \sum_{S \in \mathcal{B}} \lambda_S (c(S) + a_0 + b_0) - (c(N) + a_0 + b_0).
\end{aligned}
$$

It follows that

$$\sum_{S \in \mathcal{B}} \lambda_S \bar{c}(S) - \bar{c}(N) \geq 0$$

if and only if

$$a_0 + b_0 \geq \frac{c(N) - \sum_{S \in \mathcal{B}} \lambda_S c(S)}{\sum_{S \in \mathcal{B}} \lambda_S - 1}. \tag{5.5}$$

Let

$$q(w) = \max \frac{c(N) - \sum_{S \in \mathcal{B}} \lambda_S c(S)}{\sum_{S \in \mathcal{B}} \lambda_S - 1}.$$

where the max is taken over all minimal balanced collections. This maximum is well defined since there are only finitely many minimal balanced collections. From 5.5 it follows that $\bar{c}$ is balanced if and only if $a_0 + b_0 \geq q(w)$. □

5.3 ROUTING GAMES

Finding a travelling salesman tour which minimizes total costs is a well known $\mathcal{NP}$-hard problem. Once a tour which minimizes total costs for the grand coalition has been found, the other coalitions may well be reluctant to spend more time and resources to find optimal tours for each one of them. Let e be an optimal tour for the grand coalition. A coalition S may decide to simply use e, skipping the universities (or states, or towns, etc.) that do not belong to S, and visiting those belonging to S in the same order as e does. This induces

S	$c_e(S)$	S	$c_e(S)$	S	$c_e(S)$	S	$c_e(S)$
{1}	2.5	{1,2}	4.8	{2,4}	5.5	{1,3,4}	7.1
{2}	4.3	{1,3}	4.9	{3,4}	5.6	{2,3,4}	6.9
{3}	3.4	{1,4}	4.7	{1,2,3}	5.2	{1,2,3,4}	7.4
{4}	2.2	{2,3}	4.7	{1,2,4}	6.0	∅	0

Table 5.7 The sum of an additive game p and a routing game $\hat{c}_e$.

another cooperative cost game. Such a game we will call a *routing game*. Let $e \in E(N)$ be such that

$$w_{0e(1)} + w_{e(1)e(2)} + \cdots + w_{e(n)0} = \min_{f \in E(N)} (w_{0f(1)} + w_{f(1)f(2)} + \cdots + w_{f(n)0}).$$

In the routing game $< N, c_e >$ the cost $c_e(N)$ of the grand coalition is given by

$$c_e(N) = w_{0e(1)} + w_{e(1)e(2)} + \cdots + w_{e(n)0},$$

and the cost $c_e(S)$ of coalition S is given by

$$c_e(S) = w_{0e_S(1)} + w_{e_S(1)e_S(2)} + \cdots + w_{e_S(|S|)0}$$

where $e_S \in E(S)$ is defined by

$$e_S^{-1}(i) < e_S^{-1}(j) \Leftrightarrow e^{-1}(i) < e^{-1}(j) \text{ for all } i, j \in S.$$

Example Let us consider the non-balanced game of table 5.4. An optimal order is to do first the task of Solomon, then that of Temploy, then that of Usprint, then that of Vandam. This corresponds to $e \in E(N)$ with $e(1) = 1$, $e(2) = 2$, $e(3) = 3$, and $e(4) = 4$. The game c_e is given in table 5.7. It is the sum of the additive game p and the routing game $\hat{c}_e$. It has a non-empty core. A core element is $(1.5, 2.3, 1.4, 2.2)$. It follows that $(1,2,1,2)$ is an element of the core $C(\hat{c}_e)$ of the routing game $\hat{c}_e$.

Another optimal order is to do first the task of Vandam, then that of Solomon, then that of Temploy, and finally that of Usprint. This corresponds to $f \in E(N)$ with $f(1) = 4$, $f(2) = 1$, $f(3) = 2$, and $f(4) = 3$. The game c_f is given in table 5.8. It is the sum of the additive game p and the routing game $\hat{c}_f$. It has a non-empty core. A core element is $(2.5, 1.3, 1.4, 2.2)$. Hence, $(2,1,1,2)$ is an element of the core of the routing game $\hat{c}_f$.

In the example above we have seen that different optimal routes for N will lead to different routing games. The next theorem shows that all these games will be balanced.

S	$c_f(S)$	S	$c_f(S)$	S	$c_f(S)$	S	$c_f(S)$
{1}	2.5	{1,2}	4.8	{2,4}	4.5	{1,3,4}	7.1
{2}	4.3	{1,3}	4.9	{3,4}	3.6	{2,3,4}	4.9
{3}	3.4	{1,4}	4.7	{1,2,3}	5.2	{1,2,3,4}	7.4
{4}	2.2	{2,3}	4.7	{1,2,4}	7.0	$\emptyset$	0

Table 5.8 The sum of an additive game p and a routing game $\hat{c}_f$.

Theorem 5.3.1 *Let c_e be a routing game where $e \in E(N)$ is such that*

$$w_{0e(1)} + w_{e(1)e(2)} + \cdots + w_{e(n)0} = \min_{f \in E(N)} (w_{0f(1)} + \cdots + w_{f(n)0}).$$

Then $C(c_e) \neq \emptyset$.

Proof. Define the game $\bar{c}_e$ as follows: for each $\emptyset \neq S \subset N$, $\bar{c}_e(S)$ is equal to

$$
\begin{aligned}
\min \quad & \textstyle\sum_{i=0}^{n} \sum_{j=0}^{n} w_{ij} x_{ij} \\
\text{s.t.} \quad & \textstyle\sum_{j=0}^{n} x_{ij} = 1_S(i) && \text{for all } i \in N \\
& \textstyle\sum_{i=0}^{n} x_{ij} = 1_S(j) && \text{for all } j \in N \\
& x_{ij} \geq 0 && \text{for all } i, j \in N_0 \\
& x_{ij} = 0 && \text{for all } i, j \in S \text{ with } 1 \leq e^{-1}(j) \leq e^{-1}(i) \leq n.
\end{aligned}
$$
(5.6)

Then $\bar{c}_e$ is a linear programming game and therefore, $C(\bar{c}_e) \neq \emptyset$. Further, $\bar{c}_e(S) \leq c_e(S)$ for all $S \subset N$ because $\hat{x} \in R^{n+1} \times R^{n+1}$ defined by

$$
\hat{x}_{ij} = \begin{cases}
1 & \text{if } i, j \in S \text{ and } e_S^{-1}(i) + 1 = e_S^{-1}(j) \\
1 & \text{if } j = 0, i \in S \text{ and } e_S^{-1}(i) = |S| \\
1 & \text{if } i = 0, j \in S \text{ and } e_S^{-1}(j) = 1 \\
0 & \text{otherwise}
\end{cases}
$$

is a solution for 5.6 with $\sum_{i=0}^{n} \sum_{j=0}^{n} \hat{x}_{ij} w_{ij} = c_e(S)$. We will show that $\bar{c}_e(N) = c_e(N)$. Let us denote the solution set of 5.6 with $S = N$ by F. Let F_- be the solution set of the problem

$$
\begin{aligned}
\min \quad & \textstyle\sum_{i=0}^{n} \sum_{j=0}^{n} w_{ij} x_{ij} \\
\text{s.t.} \quad & \textstyle\sum_{j=0}^{n} x_{ij} = 1 && \text{for all } i \in N \\
& \textstyle\sum_{i=0}^{n} x_{ij} = 1 && \text{for all } i \in N \\
& x_{ij} \geq 0 && \text{for all } i, j \in N_0.
\end{aligned}
$$
(5.7)

The matrix that describes the first $2n$ constraints for F_- is totally unimodular and therefore the extreme points of F_- are integer. Also, any extreme point of

S	$c_g(S)$	S	$c_g(S)$	S	$c_g(S)$	S	$c_g(S)$
{1}	2.5	{1,2}	4.8	{2,4}	5.5	{1,3,4}	7.1
{2}	4.3	{1,3}	4.9	{3,4}	3.6	{2,3,4}	6.9
{3}	3.4	{1,4}	4.7	{1,2,3}	7.2	{1,2,3,4}	10.4
{4}	2.2	{2,3}	4.7	{1,2,4}	8.0	$\emptyset$	0

Table 5.9 The sum of the additive game p and a routing game $\hat{c}_g$ with a non-optimal tour g.

F is also an extreme point of F_-. So the extreme points of F are also integer and the minimum in 5.6 is attained at an x with $x_{ij} \in \{0,1\}$ for all $i,j \in N_0$. Due to the constraints in 5.6 such an x represents an almost travelling salesman tour, i.e., a tour in which each city, except 0 is visited exactly once, and which does not violate the order given by e. The city 0 is the only one that can be visited more than once in the tour represented by x. If this is the case, we can shorten the tour by skipping 0 every time, but the first and the last. In this way we obtain a travelling salesman tour with total costs less than or equal to $\bar{c}_e(N)$. So $c_e(N) \leq \bar{c}_e(N)$ and it follows that they are equal. Summarizing, we have

$$\bar{c}_e(N) = c_e(N)$$
$$\bar{c}_e(S) \leq c_e(S) \qquad \text{for all } S \subset N \tag{5.8}$$
$$C(\bar{c}_e) \neq \emptyset.$$

From 5.8 it follows that $C(\bar{c}_e) \subset C(c_e)$ and $C(c_e) \neq \emptyset$. $\square$

In the proof of the non-emptiness of $C(c_e)$ we really used the fact that e describes an optimal tour for N. If this is not the case then the routing game c_e need not be balanced as the following example shows.

Example Let us return to the multi-purpose machine game of table 5.4 with weights given in table 5.3. Consider the order that performs the task of Temploy first, then that of Vandam, then that of Solomon, and finally that of Usprint. This order is given by $g \in E(N)$ with $g(1) = 2$, $g(2) = 4$, $g(3) = 1$, and $g(4) = 3$. It is not an optimal order. Let c_g be the multi-purpose machine game that arises when we use this fixed order, that is, c_g is the sum of the additive game p and the routing game $\hat{c}_g$ with fixed order g. The game c_g is given in table 5.9. This game has an empty core since

$$c_g(1,2) + c_g(3,4) = 4.8 + 3.6 = 8.4 < 10.4 = c_g(N).$$

It follows that the routing game $\hat{c}_g$ has an empty core too.

Derks and Kuipers [32] showed that a routing game c_e has a non-empty core if and only if $c_e(N) \leq c_e(S) + c_e(N \backslash S)$. They describe a procedure to construct an adapted nearest neighbour tour f for which the game c_f satisfies this condition.

Routing games enable us to find another class of travelling salesman games with a non-empty core. If a travelling salesman game coincides with a routing game, we know it has a non-empty core.

Definition 5.3.2 *Let $e \in E(N)$. A coalition T is said to be e-connected if $T = \{e(i), e(i+1), \ldots, e(i+p)\}$ for some $i \in N$ and $p \geq 0$.*

We denote the set of e-connected coalitions by $\mathcal{T}(e)$.

Definition 5.3.3 *For an e-connected coalition T define the $n + 1 \times n + 1$-matrices E^T and L^T by*

$$E_{ij}^T = \begin{cases} 1 & \text{if } i \notin T \text{ and } j \in T \\ 0 & \text{otherwise} \end{cases} \qquad L_{ij}^T = \begin{cases} 1 & \text{if } i \in T \text{ and } j \notin T \\ 0 & \text{otherwise.} \end{cases}$$

Suppose the cities of T lie on an island, and the cities of $N_0 \backslash T$ lie on another island. Suppose further that travelling on the islands doesn't cost anything. Only travel between the islands has a cost attached to it. Then the matrix E^T can be viewed as describing the costs of travel to the island containing the cities of T and the matrix L^T as describing the costs of travel from the island containing the cities of T. The matrix $E^T + L^T$ gives the costs or weights for the travelling salesman problem in this situation.

Let $\mathcal{A}(e)$ be the cone generated by $\{E^T, L^T | T \in \mathcal{T}(e)\}$.

Theorem 5.3.4 *Let $e \in E(N)$ and let $W \in \mathcal{A}(e)$. Let c be the travelling salesman game with weights w_{ij} given by W. For all $S \subset N$, e_S describes an optimal tour in the game c.*

Proof. Let $W = \sum_{T \in \mathcal{T}(e)} (\alpha_T E^T + \beta_T L^T)$ with $\alpha_T, \beta_T \geq 0$. Let $S \subset N$. Then

$$w_{0 e_S(1)} + w_{e_S(1) e_S(2)} + \cdots + w_{e_S(|S|) 0} = \sum_{T \in \mathcal{T}(e)} \alpha_T (E_{0 e_S(1)}^T + \cdots + E_{e_S(|S|) 0}^T)$$

$$+ \sum_{T \in \mathcal{T}(e)} \beta_T (L_{0 e_S(1)}^T + \cdots + L_{e_S(|S|) 0}^T)$$

$$= \sum_{T \in \mathcal{T}(e):\, T \cap S \neq \emptyset} \alpha_T + \sum_{T \in \mathcal{T}(e):\, T \cap S \neq \emptyset} \beta_T.$$

Here the last equality follows from the fact that for each $T \in \mathcal{T}(e)$, the tour given by e, once it arrives in a city of T, visits every city of T before leaving T again. Let $f \in E(S)$. Then

$$w_{0f(1)} + w_{f(1)f(2)} + \cdots + w_{f(|S|)0} = \sum_{T \in \mathcal{T}(e)} \alpha_T (E^T_{0f(1)} + \cdots + E^T_{f(|S|)0})$$

$$+ \sum_{T \in \mathcal{T}(e)} \beta_T (L^T_{0f(1)} + \cdots + L^T_{f(|S|)0})$$

$$\geq \sum_{T \in \mathcal{T}(e):\, T \cap S \neq \emptyset} \alpha_T + \sum_{T \in \mathcal{T}(e):\, T \cap S \neq \emptyset} \beta_T.$$

So e_S is an optimal tour for S. $\qquad \square$

Theorem 5.3.5 *Let $e \in E(N)$ and let $W \in \mathcal{A}(e)$. Let c be the travelling salesman game with weights w_{ij} given by W. Then $C(c) \neq \emptyset$.*

Proof. From theorem 5.3.4 it follows that the travelling salesman game c and the routing game c_e coincide. So $C(c) = C(c_e) \neq \emptyset$. $\qquad \square$

Example Let $N = \{1, 2, 3, 4, 5\}$ and let $e \in e(N)$ be given by $e(1) = 1$, $e(2) = 2$, $e(3) = 3$, $e(4) = 4$, $e(5) = 5$. Let $T_1 = \{2, 3, 4\}$ and $T_2 = \{3, 4, 5\}$ Then $T_1, T_2 \in \mathcal{T}(e)$. The matrices E^{T_i} and L^{T_i} for $i = 1, 2$ are given below.

$$E^{T_1} = \begin{pmatrix} 0 & 0 & 1 & 1 & 1 & 0 \\ 0 & 0 & 1 & 1 & 1 & 0 \\ 0 & 0 & 0 & 0 & 0 & 0 \\ 0 & 0 & 0 & 0 & 0 & 0 \\ 0 & 0 & 0 & 0 & 0 & 0 \\ 0 & 0 & 1 & 1 & 1 & 0 \end{pmatrix} \quad L^{T_1} = \begin{pmatrix} 0 & 0 & 0 & 0 & 0 & 0 \\ 0 & 0 & 0 & 0 & 0 & 0 \\ 1 & 1 & 0 & 0 & 0 & 1 \\ 1 & 1 & 0 & 0 & 0 & 1 \\ 1 & 1 & 0 & 0 & 0 & 1 \\ 0 & 0 & 0 & 0 & 0 & 0 \end{pmatrix}$$

$$E^{T_2} = \begin{pmatrix} 0 & 0 & 0 & 1 & 1 & 1 \\ 0 & 0 & 0 & 1 & 1 & 1 \\ 0 & 0 & 0 & 1 & 1 & 1 \\ 0 & 0 & 0 & 0 & 0 & 0 \\ 0 & 0 & 0 & 0 & 0 & 0 \\ 0 & 0 & 0 & 0 & 0 & 0 \end{pmatrix} \quad L^{T_2} = \begin{pmatrix} 0 & 0 & 0 & 0 & 0 & 0 \\ 0 & 0 & 0 & 0 & 0 & 0 \\ 0 & 0 & 0 & 0 & 0 & 0 \\ 1 & 1 & 1 & 0 & 0 & 0 \\ 1 & 1 & 1 & 0 & 0 & 0 \\ 1 & 1 & 1 & 0 & 0 & 0 \end{pmatrix}$$

Let $\alpha_{T_1} = 2$, $\beta_{T_1} = 3$, $\alpha_{T_2} = 4$, $\beta_{T_2} = 5$. The matrix $W = \alpha_{T_1} E^{T_1} + \alpha_{T_2} E^{T_2} + \beta_{T_1} L^{T_1} + \beta_{T_2} L^{T_2}$ is given below.

$$W = \begin{pmatrix} 0 & 0 & 2 & 6 & 6 & 4 \\ 0 & 0 & 2 & 6 & 6 & 4 \\ 3 & 3 & 0 & 4 & 4 & 7 \\ 8 & 8 & 5 & 0 & 0 & 3 \\ 8 & 8 & 5 & 0 & 0 & 3 \\ 5 & 5 & 7 & 2 & 2 & 0 \end{pmatrix}$$

For the travelling salesman game c with weights w_{ij} given by W we have $c(1) = 0$, $c(2) = c(1,2) = 5$, $c(5) = c(1,5) = 9$, and $c(S) = 14$ for all other non-empty coalitions S. A core element is, for example, $(0,5,0,0,9)$. The Shapley-value $\phi(c)$ of the game is $(0, 1\frac{2}{3}, 4\frac{2}{3}, 4\frac{2}{3}, 3)$. This is also the allocation x given by

$$x_i = \sum_{T \in \mathcal{T}(e)} \left(\frac{\alpha_T}{|T|} 1_T(i) + \frac{\beta_T}{|T|} 1_T(i) \right) \text{ for all } i \in N.$$

In the following we will see that this is always the case.

Definition 5.3.6 *For each $T \subset N$ we define the game w_T by*

$$w_T(S) = \begin{cases} 1 & \text{if } S \cap T \neq \emptyset \\ 0 & \text{otherwise.} \end{cases}$$

From the proof of theorem 5.3.4 we see that

$$c = \sum_{T \in \mathcal{T}(e)} (\alpha_T w_T + \beta_T w_T). \tag{5.9}$$

Theorem 5.3.7 *Let $e \in E(N)$, $W \in \mathcal{A}(e)$, and let c be the travelling salesman game with weights w_{ij} given by W. Then c is concave.*

Proof. Immediate from 5.9 and the fact that the games w_T are concave. □

Theorem 5.3.8 (Potters) *Let c as in theorem 5.3.7. Then the Shapley-value $\phi(c)$ of c is given by*

$$\phi_i(c) = \sum_{T \in \mathcal{T}(e)} \left(\frac{\alpha_T}{|T|} 1_T(i) + \frac{\beta_T}{|T|} 1_T(i) \right).$$

Proof. From efficiency and the dummy player and symmetry properties, it follows that the Shapley-value $\phi(w_T)$ of the game w_T is given by

$$\phi_i(w_T) = \frac{1}{|T|} 1_T(i).$$

Additivity of the Shapley-value yields the desired result.   □

Potters [98] gives a larger class of matrices for which theorems 5.3.4, 5.3.5, and 5.3.8 hold. He also discusses an algorithm to determine whether a given matrix belongs to this class.

MINIMUM COST SPANNING TREE GAMES

6.1 INTRODUCTION

Four geographically separated communities in the desert, Wasteland Village, Xactustown, Yuccaville, and Zun Valley, are desperate. Their water resources are running out and they don't see a way to survive. If nothing happens they will have to leave their villages to settle somewhere else. Then, just when they think that everything is lost, a spring is found not too far away. At first there is much rejoicing, but then they have to sit down and decide on a way to build a water distribution system that is as cheap as possible. It is quite evident that to connect each community to the spring is not the cheapest way. In table 6.1 the costs involved in building a water distribution system are given in $10,000. After some analysis they realize that the cheapest thing to do is to build pipes connecting Wasteland Village and Yuccaville directly to the spring, while Zun Valley has a pipe connecting it to Wasteland Village and Xactustown has a pipe connecting it to Zun Valley. Now they have to settle on a way to distribute the costs. The total costs are 6. If they decide to divide

	Spring	Wasteland V.	Xactust.	Yuccav.	Zun V.
Spring	0	2	3	1	4
Wasteland V.	2	0	3	5	2
Xactustown	3	3	0	4	1
Yuccaville	1	5	4	0	3
Zun Valley	4	2	1	3	0

Table 6.1 The costs of building connections among the spring and the communities.

S	$c(S)$	S	$c(S)$	S	$c(S)$	S	$c(S)$
$\{1\}$	2	$\{1,2\}$	5	$\{2,4\}$	4	$\{1,3,4\}$	5
$\{2\}$	3	$\{1,3\}$	3	$\{3,4\}$	4	$\{2,3,4\}$	5
$\{3\}$	1	$\{1,4\}$	4	$\{1,2,3\}$	6	$\{1,2,3,4\}$	6
$\{4\}$	4	$\{2,3\}$	4	$\{1,2,4\}$	5	$\emptyset$	0

Table 6.2   The minimum cost spanning tree game arising from the desert water problem.

this equally, each community has to pay 1.5. Yuccaville will not agree to this because it can make a direct connection to the spring for less, namely 1. They decide to let Yuccaville pay the cost of connecting it to the spring, namely 1, let Wasteland Village do the same so it has to pay 2, let Zun Valley pay the cost of connecting it to Wasteland village, namely 2, and let Xactustown pay the cost of connecting it to Zun Valley, namely 1. To see if they can not do better in a different cooperation structure the communities decide to analyze the cooperative cost game arising from this situation. In this game it is assumed that each coalition S wants to build a water distribution system that connects only the members of S to the spring, either directly or via another member and that it wants to do this as cheaply as possible. This game is given in table 6.2. Here Wasteland Village=1, Xactustown=2, Yuccaville=3, and Zun Valley=4. It is easy to verify that the cost allocation (2,1,1,2) prescribed above, is an element of the core of this game. It is, in fact, an extreme point of the core of the game.

The problem of finding a cheapest way to build a water distribution system servicing all the four communities can be formulated as the problem to find a *minimum cost spanning tree* for the complete graph on 5 nodes with the costs of the edges given in table 6.1. For each coalition S, $c(S)$ is determined by the cost of a minimum cost spanning tree for the complete graph on the nodes of S plus the spring. Such a game is called a *minimum cost spanning tree game*. We will use the abbreviation *mcst-game*.

The problem of allocating the costs of a spanning tree in a graph among the users which are situated at the nodes of a graph, with one node reserved for a common supplier which is not to participate in the cost sharing, has been first introduced by Claus and Kleitman [13]. Bird [8] was the first to suggest a game theoretic approach to the problem. He also proposed the cost allocation that we used in the example. Granot and Huberman [57] showed that this cost allocation will always be in the core of the mcst-game. In the subsequent sections we will study mcst-games in more detail. In section 6.2 we will give the formal definition of these games and of the solution proposed by Bird. We will also discuss an axiomatic characterization of this solution given by

Feltkamp et al. [46]. In section 6.3 we will study a subset of the core of a mcst-game, the irreducible core introduced by Bird [8]. The Shapley-value of a mcst-game need not be an element of the core. In section 6.4 we will look at the restricted weighted Shapley-value introduced by Bird. The restricted weighted Shapley-value of a mcst-game always lies in the core of the game. In that section we will also discuss permutationally concave games, a generalization of concave games, introduced by Granot and Huberman [58]. Every mcst-game is a permutationally concave game. Quite some work has been done on games related to mcst-games and in section 6.5 we will briefly discuss some of this work.

6.2 MINIMUM COST SPANNING TREE GAMES

Let $N = \{1, 2, \ldots, n\}$ be a set of geographically separated users of some good that is supplied by a common supplier 0, and for which a distribution system consisting of links among the members of $N_0 = N \cup \{0\}$ has to be built. There is a cost associated with the building of each link. Let $G = (N_0, E)$ be the complete graph with set of nodes N_0 and set of edges denoted by E. Let $k_{ij} = k_{ji}$ denote the cost of constructing the link $\{i, j\} \in E$. A *tree* is a connected graph which contains no cycles. A *spanning tree* for a given connected graph is a tree, with set of nodes equal to the set of nodes of the given graph, and set of edges a subset of the set of edges of the given graph. A *minimum cost spanning tree* for a given connected graph with costs on the edges, is a spanning tree which has least cost among all spanning trees.

Definition 6.2.1 *The cost $c(S)$ of coalition $S \subset N$ in a minimum cost spanning tree game c is equal to*

$$\sum_{\{i,j\} \in E_{T_S}} k_{ij}$$

where E_{T_S} is the set of edges of a minimum cost spanning tree in the complete graph $G_S = (S_0, E_S)$. Here $S_0 = S \cup \{0\}$.

Two well-known algorithms for finding a minimum cost spanning tree in a given graph are the Prim-Dijkstra [102, 33] algorithm and the Kruskal [75] algorithm. The Prim-Dijkstra algorithm starts with an arbitrary node and builds a connected graph by every time choosing a cheapest edge, to make a connection between the part of the tree that has already been constructed, and the nodes that are not yet connected. In this way all minimum cost spanning trees of a graph can be found. Kruskal's algorithm selects edges to belong to the tree to be constructed, by starting with a cheapest edge and adding each time a cheapest edge among the ones that have not been selected yet, which does not create a cycle. It will also generate all minimum cost spanning trees. Let c be a mcst-game and let $E_T \subset E$ be the set of edges of a minimum cost spanning tree T for the graph (N_0, E). Bird [8] proposed the following cost allocation scheme. For each $i \in N$ let the amount that i has to pay be equal to the cost of the edge incident upon i on the unique path from 0 to i in T. It is easy to see that in this way the total costs are distributed among the players. Since there can be more than one minimum cost spanning tree for a graph, this way of dividing the costs need not lead to a unique cost allocation in a mcst-game. We will call any allocation that arises in this way a *Bird tree allocation*.

Theorem 6.2.2 *Let c be a mcst-game. Let x be a Bird tree allocation for c. Then $x \in C(c)$.*

Proof. Let T with set of edges E_T, be the minimum cost spanning tree that defines x. It is clear that $x(N) = c(N)$. For $S \subset N$ let T_S be a minimum cost spanning tree on the graph (S_0, E_S). Construct a spanning tree $\hat{T}$ for (N_0, E) as follows. Add all the nodes in $N \setminus S$ to T_S and for each $i \in N \setminus S$ add the edge incident upon i on the unique path from 0 to i in T to E_S. Then $\hat{T}$ constructed in this way is a spanning tree for (N_0, E). So

$$c(S) + x(N \setminus S) = \sum_{\{i,j\} \in E_{\hat{T}}} k_{ij} \geq \sum_{\{i,j\} \in E_T} k_{ij} = c(N),$$

and it follows that $c(S) \geq x(S)$.   □

As the following example shows a mcst-game need not be monotonic.

Example For the mcst-game corresponding to figure 6.1 we have $c(3) = 6 > 4 = c(1,3)$. We can define the *monotonic cover* $\bar{c}$ of the game c by making $\bar{c}(S)$ equal to the cost of any tree in the graph that has a set of nodes that contains S_0. The mcst-game c and its monotonic cover $\bar{c}$ are given in table 6.3. Neither of the two games is concave.

Since $\bar{c}(S) \leq c(S)$ for all $S \subset N$ and $\bar{c}(N) = c(N)$ for a mcst-game c and

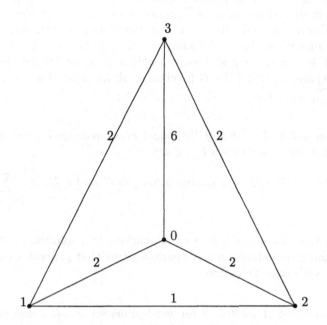

Figure 6.1 A graph leading to a non-monotonic mcst-game.

S	$c(S)$	$\bar{c}(S)$	S	$c(S)$	$\bar{c}(S)$	S	$c(S)$	$\bar{c}(S)$
$\{1\}$	2	2	$\{1,2\}$	3	3	$\{1,2,3\}$	5	5
$\{2\}$	2	2	$\{1,3\}$	4	4	$\emptyset$	0	0
$\{3\}$	6	4	$\{2,3\}$	4	4			

Table 6.3 A mcst-game c and its monotonic cover $\bar{c}$.

its monotonic cover $\bar{c}$ it follows that $C(\bar{c}) \subset C(c)$. Similarly to the proof of theorem 6.2.2 one can show that a Bird tree allocation is in the core of $C(\bar{c})$. In fact, it is an extreme point of $C(c)$ and therefore also of $C(\bar{c})$.
In the following we will discuss an axiomatic characterization of the set of Bird tree allocations due to Feltkamp et al. [46]. Given a graph $G = (N_0, E)$ with costs k_{ij} on the edges, a *mcst problem* is defined to be the problem of finding a spanning tree T for G together with an allocation $x \in R^n$ satisfying $x(N) \geq \sum_{\{i,j\} \in E_T} k_{ij}$.

Definition 6.2.3 *A solution f for a mcst-problem assigns to every mcst-problem given by a graph G with costs k_{ij}, a set*

$$f(G, k) \subset \{(T, x) | T \text{ is a spanning tree for } G \text{ and } x(N) \geq \sum_{\{i,j\} \in E_T} k_{ij}\}.$$

It is clear that the set of Bird tree allocations is a solution as defined above if we combine the minimum cost spanning tree that generates each Bird tree allocation with that allocation.

Definition 6.2.4 *A solution f for mcst-problems is called non-empty if $f(G, k) \neq \emptyset$ for all mcst-problems (G, k).*

Definition 6.2.5 *A solution f for mcst-problems is called efficient if for all mcst-problems (G, k) the following holds: For all $(T, x) \in f(G, k)$*

(a) *T is a minimum cost spanning tree for (G, k).*

(b) *$x(N) = \sum_{\{i,j\} \in E_T} k_{ij}$.*

It is evident that the set of Bird tree allocations is non-empty and efficient.
A player is called a *leaf* in a graph if there is only one edge incident upon this player. Let $(T, x) \in f(G, k)$ and let $i \in N$ be a leaf in T. Then no other player makes use of i to be connected to 0, so if we delete i and all of the edges incident upon i from G, this should not effect the other players. This is reflected in the so-called *leaf consistency property*.

Definition 6.2.6 *Let (G, k) be a mcst-problem and let $i \in N$. The reduced mcst-problem is the mcst-problem (G^{-i}, k^{-i}). Here G^{-i} is the graph from which* *i* and all the edges incident upon *i* have been removed, and k^{-i} is the restriction of *k* to the edge set E^{-i} of G^{-i}.

Definition 6.2.7 *A solution f for mcst-problems is said to be leaf consistent if for all mcst-problems (G, k) the following holds: if $(T, x) \in f(G, k)$ and i is a leaf in T then $(T^{-i}, x^{-i}) \in f(G^{-i}, k^{-i})$. Here T^{-i} is obtained by deleting i and the edge incident upon i, and x^{-i} is obtained by removing x_i from x.*

Definition 6.2.8 *A solution f for mcst-problems is called converse leaf consistent if for all mcst-problems (G, k) the following holds: If T is a minimum cost spanning tree for G and x satisfies $x(N) = \sum_{\{i,j\} \in E_T} k_{ij}$ then*

$$(T^{-i}, x^{-i}) \in f(G^{-i}, k^{-i}) \text{ for all i that are leaves of } T$$

implies

$$(T, x) \in f(G, k).$$

Converse leaf consistency implies that if an efficient solution is found suitable after deleting a leaf, it should also be suitable for the situation containing the leaf.

Theorem 6.2.9 *The set of Bird tree allocations is the unique non-empty solution for mcst-problems that satisfies efficiency, leaf consistency, and converse leaf consistency.*

Proof. As was mentioned before, the set of Bird tree allocations is a non-empty, efficient solution for mcst-problems. Let (G, k) be a mcst-problem. Suppose (T, x) is in the set of Bird tree allocations and suppose *i* is a leaf in *T*. Then T^{-i} is a minimum cost spanning tree for G^{-i} since *T* is a minimum cost spanning tree for *G* and *i* is a leaf in *T*. Also $x(N \setminus \{i\}) = \sum_{\{j,l\} \in E_{T^{-i}}} k_{jl}$ by definition of the Bird cost allocation. So (T^{-i}, x^{-i}) is an element of the set of Bird tree allocations and therefore this set satisfies leaf consistency.

Suppose (T, x) is such that *T* is a minimum cost spanning tree for *G* and $x(N) = \sum_{\{i,j\} \in E_T} k_{ij}$. Suppose further that (T^{-i}, x^{-i}) is an element of the set of Bird tree allocations for (G^{-i}, k^{-i}) for any player *i* that is a leaf in *T*. Then x_i is equal to the cost of the edge incident upon *i* in *T*. So (T, x) is an element of the set of Bird tree allocations.

Let *f* be a non-empty solution for mcst-problems which satisfies efficiency, leaf consistency, and converse leaf consistency. We will use induction on the number of players $|N|$ to show that f is equal to the set of Bird tree allocations. For $|N| = 1$ there is exactly one efficient solution so f and the set of Bird tree allocations coincide. Suppose that they coincide for all mcst-problems with $|N| \leq k$. Let (G, k) be a mcst-problem with $|N| = k + 1$. Suppose $(T, x) \in f(G, k)$. Let $i \in N$ be a leaf in T. Such an i always exists because T is a tree for G. Then $(T^{-i}, x^{-i}) \in f(G^{-i}, k^{-i})$ because of leaf consistency. Because of the induction assumption it follows that (T^{-i}, x^{-i}) is an element of the set of Bird tree allocations for (G^{-i}, k^{-i}). Since this is true for any leaf i in T we obtain from converse leaf consistency that (T, x) is an element of the set of Bird tree allocations for (G, k). So $f(G, k)$ is a subset of the set of Bird tree allocations for (G, k). Similarly one can show that any Bird tree allocation for (G, k) will also be an element of $f(G, k)$. So $f(G, k)$ coincides with the set of Bird tree allocations for (G, k).   □

6.3   THE IRREDUCIBLE CORE

The *irreducible core* of a mcst-game was introduced by Bird [8] as a means of generating more core allocations over those given in the set of Bird tree allocations. This can be done if we have a minimum cost spanning tree for the mcst-problem that we are considering, without any knowledge of the costs of the edges not in this tree.

Definition 6.3.1 *Let c be a mcst-game arising from the mcst-problem (G, k) with a minimum cost spanning tree T. We define a new mcst-problem (G, k^T) by*

$$
k^T_{ij} := \begin{cases} k_{ij} & \text{if } \{i, j\} \in E_T \\ max \; \{k_{hl} | \{h, l\} \text{ an edge on the unique path} \\ \qquad \text{connecting } i \text{ and } j \text{ in } T\} & \text{otherwise.} \end{cases}
$$

We denote by c^T the mcst-game generated by (G, k^T).
It is easy to verify that $k^T_{ij} \leq k_{ij}$ for all $\{i, j\} \in E$ and that T is a minimum cost spanning tree in the mcst-problem (G, k^T).

Example Let (G, k) be as given in figure 6.1 and let T be the minimum cost

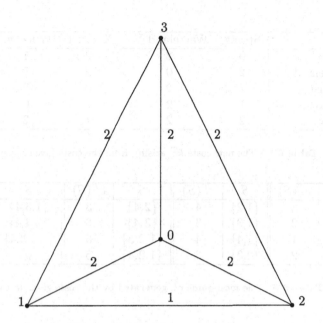

Figure 6.2 The graph with costs (G, k^T) arising from the graph of figure 6.1.

S	$c^T(S)$	S	$c^T(S)$	S	$c^T(S)$	S	$c^T(S)$
$\{1\}$	2	$\{3\}$	2	$\{1,3\}$	4	$\{1,2,3\}$	5
$\{2\}$	2	$\{1,2\}$	3	$\{2,3\}$	4	$\emptyset$	0

Table 6.4 The mcst-game c^T generated by (G, k^T) as given in figure 6.2.

spanning tree with $E_T = \{\{0,1\}, \{1,2\}, \{2,3\}\}$. The new graph with costs (G, k^T) is given in figure 6.2. The game c^T generated by (G, k^T) is given in table 6.4.

Example Consider the mcst-game of table 6.2 with costs given in table 6.1. Let T be the minimum cost spanning tree with $E_T = \{\{0,1\}, \{0,3\}, \{1,4\}, \{2,4\}\}$. The new costs k^T are given in table 6.5. The mcst-game c^T generated by these new costs is given in table 6.6.

Proposition 6.3.2 *Let (G, k) be a mcst-problem with a minimum cost spanning tree T and let (G, k^T) be as defined in definition 6.3.1. Let T' be another minimum cost spanning tree for (G, k). Then T' is also a minimum cost spanning tree for (G, k^T) and $k_{ij}^T = k_{ij}^{T'}$ for all $\{i, j\} \in E$.*

	Spring	Wasteland V.	Xactust.	Yuccav.	Zun V.
Spring	0	2	2	1	2
Wasteland V.	2	0	2	2	2
Xactust.	2	2	0	2	1
Yuccav.	1	2	2	0	2
Zun V.	2	2	1	2	0

Table 6.5 The new costs k^T arising from the costs given in table 6.1

S	$c^T(S)$	S	$c^T(S)$	S	$c^T(S)$	S	$c^T(S)$
$\{1\}$	2	$\{1,2\}$	4	$\{2,4\}$	3	$\{1,3,4\}$	5
$\{2\}$	2	$\{1,3\}$	3	$\{3,4\}$	3	$\{2,3,4\}$	4
$\{3\}$	1	$\{1,4\}$	4	$\{1,2,3\}$	5	$\{1,2,3,4\}$	6
$\{4\}$	2	$\{2,3\}$	3	$\{1,2,4\}$	5	$\emptyset$	0

Table 6.6 The mcst-game c^T generated by the costs given in table 6.5.

Proof. Let T' be a minimum cost spanning tree for (G, k). Then

$$\sum_{\{i,j\}\in E_{T'}} k^T_{ij} \leq \sum_{\{i,j\}\in E_{T'}} k_{ij} = \sum_{\{i,j\}\in E_T} k_{ij} = \sum_{\{i,j\}\in E_T} k^T_{ij}. \qquad (6.1)$$

From 6.1 and the fact that T is a minimum cost spanning tree for (G, k^T) it follows that the inequality in 6.1 is an equality and that T' is a minimum cost spanning tree for (G, k^T). Hence $k^T_{ij} = k_{ij}$ for all $\{i,j\} \in E_{T'}$. Also, by applying the procedure described in definition 6.3.1 on (G, k^T) with T' as the minimum cost spanning tree we obtain

$$k^{T'}_{ij} = (k^T)^{T'}_{ij} \leq k^T_{ij} \text{ for all } \{i,j\} \in E.$$

Similarly, by switching the roles of T and T' we find that

$$k^T_{ij} \leq k^{T'}_{ij} \text{ for all } \{i,j\} \in E.$$

So $k^T_{ij} = k^{T'}_{ij}$ for all $\{i,j\} \in E$.   □

From proposition 6.3.2 it follows that it doesn't matter which minimum cost spanning tree we take in (G, k), they will all lead to the same new cost function on the edges. So we can use the notation $(G, \hat{k})$ instead of (G, k^T), and $\hat{c}$ instead of c^T for the corresponding mcst-game.

Definition 6.3.3 *Let c be a mcst-game arising from* (G, k). *The irreducible core* $IC(c)$ *of c is defined to be the core* $C(\hat{c})$ *of* $\hat{c}$.

Example The irreducible core of the mcst-game of table 6.3 is the core of the game $\hat{c}$, which is the same as the game c^T of table 6.4. So $IC(c) = C(\hat{c})$ is equal to the convex hull of the points (1,2,2) and (2,1,2), whereas the core of the mcst-game c is equal to the convex hull of the points (1,2,2), (2,1,2), and (1,1,3).

The irreducible core of the mcst-game given in table 6.2 is the core of the game given in table 6.6 which is equal to the convex hull of the points (2,1,1,2) and (2,2,1,1), whereas the core of the mcst-game of table 6.2 is equal to the convex hull of the points (2,3,1,0), (2,1,1,2), (1,3,1,1), and (1,1,1,3).

Since $\hat{c}(N) = c(N)$ and $\hat{c}(S) \leq c(S)$ for all $S \subset N$ it follows that $IC(c) \subset C(c)$. Bird [8] gave a nice characterization of $IC(c)$. Granot and Huberman showed that $\hat{c}$ is concave. To prove these results we will need the following lemma.

Lemma 6.3.4 *Let* (G, k) *be a mcst-problem. Let* $(G, \hat{k})$ *be the mcst-problem derived form* (G, k) *by the procedure in definition 6.3.1. Let* $S \subset N$.

(a) *If* T_S *is a minimum cost spanning tree for* $(G_S, \hat{k})$ *then* $(\hat{k})_{ij}^{T_S} = \hat{k}_{ij}$ *for all* $\{i, j\} \in E_S$.

(b) *For all* $i \in S$ *there exists a minimum cost spanning tree* T_S^i *such that* i *is a leaf in* T_S^i.

(c) *Let* T_S *be a minimum cost spanning tree for* $(G_S, \hat{k})$ *and let* $i \in N \setminus S$. *Let* $j \in S_0$ *be such that*
$$\hat{k}_{ij} = \min\{\hat{k}_{il} | l \in S_0\}.$$
Then the tree $T_{S \cup \{i\}}$ *with* $E_{T_{S \cup \{i\}}} = E_{T_S} \cup \{i, j\}$ *is a minimum cost spanning tree for* $(G_{S \cup \{i\}}, \hat{k})$.

Proof. (a) By the definition of $(\hat{k})_{ij}^{T_S}$ we know that it is less than or equal to $\hat{k}_{ij}$. We have to prove the inverse inequality. Let $\{i, j\} \in E_S$. Let T be a minimum cost spanning tree for (G, k) and let $\{l, m\}$ be a maximum cost edge on the path connecting i and j in T. By deleting this edge from T we will remain with two components, one containing i and the other containing j. Let $\{o, q\}$ be an edge on the path connecting i and j in T_S such that it links these two components. Then $\{l, m\}$ lies on the path that connects o and q in T, so $\hat{k}_{oq} \geq k_{lm}$. This yields

$$\hat{k}_{ij} = k_{lm} \leq \hat{k}_{oq}.$$

Since $\{o, q\}$ lies on the path connecting i and j in T_S it follows that $\hat{k}_{ij} \leq (\hat{k})_{ij}^{T_S}$.

(b) Let $i \in S$ and let T_S be a minimum cost spanning tree for $(G_S, \hat{k})$. If i is a leaf in T_S we are done. Suppose i is not a leaf in T_S, say there are $\delta > 1$ edges incident upon i. Let $p(i)$ be the immediate predecessor of i on the path from 0 to i in T_S. Let j be such that i is the immediate predecessor of j on the path connecting 0 with j in T_S. Then it follows from (a) that

$$\hat{k}_{p(i)j} = (\hat{k}_{p(i)j})^{T_S} = \max\{\hat{k}_{p(i)i}, \hat{k}_{ij}\}.$$

Let us construct a minimum cost spanning tree T_S' for $(G_S, \hat{k})$ by setting

$$E_{T_S'} = \left\{ \begin{array}{ll} E_{T_S} \setminus \{\{p(i), i\}\} \cup \{\{p(i), j\}\} & \text{if } \hat{k}_{p(i)j} = \hat{k}_{p(i)i} \\ E_{T_S} \setminus \{\{i, j\}\} \cup \{\{p(i), j\}\} & \text{if } \hat{k}_{p(i)j} = \hat{k}_{ij}. \end{array} \right.$$

Note that T_S' is a minimum cost spanning tree for $(G_S, \hat{k})$ in which we have decreased the number of edges incident upon i to $\delta - 1$. By continuing this process, if necessary, we will arrive at a minimum cost spanning tree for $(G_S, \hat{k})$ in which i is a leaf.

(c) Let $\bar{T}_{S \cup \{i\}}$ be a minimum cost spanning tree for $(G_{S \cup \{i\}}, \hat{k})$ with i being a leaf in it. Let $\pi(i)$ be the immediate predecessor of i in $\bar{T}_{S \cup \{i\}}$. Because i is a leaf, we know that

$$\hat{k}_{\pi(i)i} = \min\{\hat{k}_{il} | l \in S_0\} = \hat{k}_{ij}.$$

Also, we know that $\bar{T}_S$ given by $E_{\bar{T}_S} = E_{\bar{T}_{S \cup \{i\}}} \setminus \{\{\pi(i)i\}\}$ is a minimum cost spanning tree for $(G_S, \hat{k})$. Therefore,

$$\sum_{\{j,l\} \in E_{T_{S \cup \{i\}}}} \hat{k}_{jl} = \sum_{\{j,l\} \in E_{T_S}} \hat{k}_{jl} + \hat{k}_{ij} = \sum_{\{j,l\} \in E_{\bar{T}_S}} \hat{k}_{jl} + \hat{k}_{ij} = \sum_{\{j,l\} \in E_{\bar{T}_{S \cup \{i\}}}} \hat{k}_{jl}.$$

Since $\bar{T}_{S \cup \{i\}}$ is a minimum cost spanning tree for $(G_{S \cup \{i\}}, \hat{k})$ it follows that $T_{S \cup \{i\}}$ is also a minimum cost spanning tree for $(G_{S \cup \{i\}}, \hat{k})$. $\qquad\square$

Lemma 6.3.4 states that, given a $(G, \hat{k})$ and a $S \subset N$, (a) repeating the procedure given in definition 6.3.1 on $(G_S, \hat{k})$ will not change anything, (b) for every member of S there exists a minimum cost spanning tree for $(G_S, \hat{k})$ in which that member is a leaf, and (c) given a minimum cost spanning tree for $(G_S, \hat{k})$ one can extend it to include a non-member of S by using a cheapest edge to make the connection between the non-member and S.

Theorem 6.3.5 (Granot and Huberman) *Let* (G, k) *be a mcst-problem. Let* $(G, \hat{k})$ *be the mcst-problem arising from* (G, k) *by the procedure of definition 6.3.1. Let* $\hat{c}$ *be the mcst-game generated by* $(G, \hat{k})$. *Then* $\hat{c}$ *is concave.*

Proof. Let $S^2 \subset S^1 \subset N \setminus \{i\}$. Then we obtain with the aid of lemma 6.3.4 that

$$
\begin{aligned}
\hat{c}(S^2 \cup \{i\}) - \hat{c}(S^2) &= \min\{\hat{k}_{ij} | j \in S_0^2\} \\
&\geq \min\{\hat{k}_{ij} | j \in S_0^1\} \\
&= \hat{c}(S^1 \cup \{i\}) - \hat{c}(S^1).
\end{aligned}
$$

So $\hat{c}$ is concave. $\qquad\square$

Theorem 6.3.6 (Bird) *Let* $\hat{c}$ *be as in theorem 6.3.5 and let* c *be the mcst-game arising from* (G, k), *then* $IC(c) = C(\hat{c})$ *=the convex hull of the set of Bird tree allocations for* $(G, \hat{k})$.

Proof. From theorem 6.2.2 it follows that every Bird tree allocation for $(G, \hat{k})$ is an element of the core of $\hat{c}$. Because the core is a convex set it follows that the convex hull of the set of Bird tree allocations for $(G, \hat{k})$ is a subset of $C(\hat{c})$. We need to show the inverse inclusion. Since $\hat{c}$ is a concave cost game we know that the core is the convex hull of the marginal vectors $m^\sigma(\hat{c})$. Given such a marginal vector $x = m^\sigma(\hat{c})$ we will construct a minimum cost spanning tree for $(G, \hat{k})$ such that x is the Bird tree allocation corresponding to that tree. Let us define the coalitions $S^\sigma(m)$ for all $m \in N_0$ by

$$
\begin{aligned}
S^\sigma(0) &:= \emptyset \\
S^\sigma(m) &:= \{\sigma^{-1}(j) | 1 \leq j \leq m\} \quad \text{for all } m \in N.
\end{aligned}
$$

Then the set of predecessors of $\sigma^{-1}(i)$ in the order given by σ

$$
P(\sigma, \sigma^{-1}(i)) = \{j \in N | \sigma(j) < i\} = S^\sigma(i - 1).
$$

We use the following iterative procedure to construct the desired minimum cost spanning tree. Let T_1^x be the minimum cost spanning tree for $(G_{S^\sigma(1)}, \hat{k})$. Given a minimum cost spanning tree T_{i-1}^x for $(G_{S^\sigma(i-1)}, \hat{k})$ we extend it to a minimum cost spanning tree T_i^x for $(G_{S^\sigma(i)}, \hat{k})$ by connecting $\sigma^{-1}(i)$ to T_{i-1}^x with a cheapest edge as described in (c) of lemma 6.3.4. The tree T_n^x is a minimum cost spanning tree for $(G, \hat{k})$. Let $p(\sigma^{-1}(i))$ be the immediate predecessor of $\sigma^{-1}(i)$ in T_n^x. Since

$$\hat{c}(S^\sigma(m)) = \sum_{\{j,l\} \in E_{T_m^x}} \hat{k}_{jl} \text{ for all } m \in N,$$

it follows that

$$\begin{aligned} x_{\sigma^{-1}(i)} &= \hat{c}(S^\sigma(i-1) \cup \{\sigma^{-1}(i)\}) - \hat{c}(S^\sigma(i-1)) \\ &= \hat{k}_{p(\sigma^{-1}(i)i)}. \end{aligned}$$

Hence x is the Bird tree allocation generated by the minimum cost spanning tree T_n^x.   □

Given two mcst-games c^1 and c^2 arising from (G^1, k^1) and (G^2, k^2), respectively, if T is a minimum cost spanning tree in both (G^1, k^1) and (G^2, k^2) and $k_{ij}^1 = k_{ij}^2$ for all $\{i,j\} \in E_T$ then $IC(c^1) = IC(c^2)$. So the irreducible core is a subset of the core for all mcst-games that have a minimum cost spanning tree with fixed costs on the edges in common.

Bird [8] and Aarts [1] give conditions that guarantee that the irreducible core will have exactly one element.

6.4 THE RESTRICTED WEIGHTED SHAPLEY-VALUE AND PERMUTATIONALLY CONCAVE GAMES

As the mcst-game from table 6.3 illustrates the Shapley-value need not be in the core of a mcst-game. For that game the Shapley-value is $(\frac{5}{6}, \frac{5}{6}, \frac{20}{6})$ and it is easy to verify that this is not in the core. Bird remarks that this is due to the fact that the Shapley-value considers orders of coalition formation that would not occur if only minimum cost spanning trees were used. To remedy this he proposes a *restricted weighted Shapley-value* which considers only orders that are defined as being feasible.

Definition 6.4.1 *Let c be a mcst-game arising from the mcst-problem (G, k). A permutation $\pi \in \Pi_N$ is said to be feasible for c if there exists a minimum cost spanning tree T for (G, k) such that for each $1 \le p \le n$, T contains a minimum cost spanning tree for $(G_{S^\pi(p)}, k)$. Here $S^\pi(p) = \{\pi^{-1}(1), \pi^{-1}(2), \dots, \pi^{-1}(p)\}$.*

Let $F\Pi_N$ denote the set of feasible permutations of N. The set of restricted weighted Shapley-values is defined by considering only permutations that are in $F\Pi_N$. Let us denote the set of restricted weighted Shapley-values for the mcst-game c by $\rho\phi(c)$. Then

$$\rho\phi(c) := \{ \sum_{\pi \in F\Pi_N} p_\pi m^\pi(c) | \sum_{\pi \in F\Pi_N} p_\pi = 1, \ p_\pi \geq 0 \text{ for all } \pi \in F\Pi_N \}.$$

Theorem 6.4.2 (Bird) *Let c be a mcst-game. Then $\rho\phi(c) \subset IC(c)$.*

Proof. Let π be a feasible permutation. Then $m^\pi(c)$ is a Bird tree allocation for c and therefore also for $\hat{c}$. So any convex combination of the marginal vectors corresponding to feasible permutations is in $IC(c)$. □

Example For the mcst-game in table 6.3 the feasible permutations correspond to the orders: 123, 132, 213, 231. The first two correspond to the Bird tree allocation (2,1,2), and the last two to the Bird tree allocation (1,2,2). Any convex combination of these two will be in the irreducible core as well as in the set of restricted weighted Shapley-values. In this example these two sets coincide.

Example For the mcst-game in table 6.2 the feasible permutations correspond to the orders: 1423, 1432, 1342, 3142. All four correspond to the same Bird tree allocation (2,1,1,2) so this is the unique restricted weighted Shapley-value for this game. Recall that the irreducible core is the convex hull of (2,1,1,2) and (2,2,1,1).

Granot and Huberman [58] introduced permutationally concave games. Recall that one of the characterizations of a concave game states that a game c is concave if and only if for all $i \in N$ and $S^2 \subset S^1 \subset N \setminus R$ the following holds

$$c(S^1 \cup R) - c(S^1) \leq c(S^2 \cup R) - c(S^2). \tag{6.2}$$

The definition of permutational concavity is related to this.

Definition 6.4.3 *A cooperative cost game c is called permutationally concave if there exists a permutation $\pi \in \Pi_N$ such that for all $1 \leq p_2 \leq p_1 \leq n$ and all $R \subset N \setminus S^\pi(p_1)$ the following holds*

$$c(S^\pi(p_1) \cup R) - c(S^\pi(p_1)) \leq c(S^\pi(p_2) \cup R) - c(S^\pi(p_2)). \tag{6.3}$$

Note that for a game to be permutationally concave, 6.2 is required to hold only for certain coalitions given by a permutation π.

Theorem 6.4.4 (Granot and Huberman) *Let c be a permutationally concave cost game. Then $C(c) \neq \emptyset$.*

Proof. Let π be a permutation satisfying 6.3. We will show that the marginal vector $m^\pi(c)$ is in the core of c. It is clear that $\sum_{i \in N} m_i^\pi(c) = c(N)$. Let $\emptyset \neq S \subset N$, say $S = \{i_1, i_2, \ldots, i_k\}$ with $\pi(i_1) \leq \pi(i_2) \leq \ldots \leq \pi(i_k)$. We will use the notation $\overline{P(j)} = P(\pi, j) \cup \{j\}$. Then using 6.3 repeatedly we obtain

$$
\begin{aligned}
\sum_{j \in S} m_j^\pi(c) &= \sum_{r=1}^{k} m_{i_r}^\pi(c) \\
&= \sum_{r=1}^{k} [c(\overline{P(i_r)}) - c(P(\pi, i_r))] \\
&= c(\overline{P(i_k)}) - c(P(\pi, i_k)) + \sum_{r=1}^{k-1} [c(\overline{P(i_r)}) - c(P(\pi, i_r))] \\
&\leq c(\overline{P(i_{k-1})} \cup \{i_k\}) - c(\overline{P(i_{k-1})}) + \sum_{r=1}^{k-1} [c(\overline{P(i_r)}) - c(P(\pi, i_r))] \\
&= c(\overline{P(i_{k-1})} \cup \{i_k\}) - c(P(\pi, i_{k-1})) + \sum_{r=1}^{k-2} [c(\overline{P(i_r)}) - c(P(\pi, i_r))] \\
&\leq c(\overline{P(i_{k-2})} \cup \{i_k, i_{k-1}\}) - c(\overline{P(i_{k-2})}) \\
&\quad + \sum_{r=1}^{k-2} [c(\overline{P(i_r)}) - c(P(\pi, i_r))] \\
&= c(\overline{P(i_{k-2})} \cup \{i_k, i_{k-1}\}) - c(P(\pi, i_{k-2})) \\
&\quad + \sum_{r=1}^{k-3} [c(\overline{P(i_r)}) - c(P(\pi, i_r))].
\end{aligned}
$$

Continuing in this fashion we arrive at

$$
\sum_{j \in S} m_j^\pi(c) \leq c(\overline{P(i_1)} \cup \{i_k, i_{k-1}, \ldots, i_2\}) - c(P(\pi, i_1))
$$

$$= c(P(\pi, i_1) \cup \{i_k, i_{k-1}, \ldots, i_2, i_1\}) - c(P(\pi, i_1))$$
$$\leq c(i_k, i_{k-1}, \ldots, i_2, i_1) = c(S)$$

where the last inequality follows from subadditivity. So $m^\pi(c) \in C(c)$. $\qquad\square$

Example The mcst-games from tables 6.2 and 6.3 are both permutationally concave. A permutation satisfying 6.3 for the first one is given by the order 1423, and for the second one by 123.

Theorem 6.4.5 (Granot and Huberman) *Let c be a mcst-game. Then c is permutationally concave.*

Proof. Let c be a mcst-game arising from the mcst-problem (G, k). Let T be a minimum cost spanning tree for (G, k). Let π be any permutation satisfying:

For all $i, j \in N$, if i is on the unique path from 0 to j in T then $\pi(i) < \pi(j)$.

Then 6.3 is satisfied with π. $\qquad\square$

By reversing the inequality in 6.3 we obtain the definition of a *permutationally convex game*. The result of theorem 6.4.4 also holds for revenue games that are permutationally convex.

The result of Bird on restricted weighted Shapley-values and the result of Granot and Huberman on permutationally concave games are related to each other. Let c be a mcst-game. Inequality 6.3 is satisfied by each feasible permutation. Further, for every permutationally concave game c, if π is a permutation for which 6.3 holds, the marginal vector $m^\pi(c)$ is in the core of c. Hence, every convex combination of such marginal vectors is in the core of c. This yields another proof of the fact that the set of restricted weighted Shapley-values is a subset of the core of a mcst-game. It also shows that the definition of restricted weighted Shapley-values can be extended to all permutationally concave games, by considering all permutations for which 6.3 holds as feasible.

6.5 GENERALIZATIONS, VARIATIONS, AND SPECIAL CASES

A situation where the players can use other links besides those connecting two customers or a customer with the supplier is studied by Megiddo [82] He shows that the cooperative game arising from such a situation can have an empty core.

In [83] Megiddo studies games in which a spanning tree T of $G = (N_0, k)$ is given and the cost of a coalition S is the total cost of the edges that belong to some path from 0 to a node of S. In a straightforward way it can be proved that these games are concave. Megiddo gives polynomial algorithms to compute the nucleolus and Shapley-value of these games.

As mentioned in section 2 a Bird tree allocation is an extreme point of the core of a mcst-game. Granot and Huberman [59] claim that a Bird tree allocation discriminate against players that are closest to the root, making them pay all the costs of connecting to the root, while players that are farther away also benefit form this connection. To remedy this they consider two operations that allow players, that are closer to the root, to deflect the cost allocated to them, to players that are connected to the root via them. These are the so-called *weak demand* and *strong demand* operations. When applied to a Bird tree allocation these operations yield more core elements.

In the same paper Granot and Huberman showed that the nucleolus of a mcst-game is the unique point in the intersection of the core and the kernel. They also showed that if T is a minimum cost spanning tree for the mcst-problem (G, k) then the nucleolus of the corresponding mcst-game c depends only on the coalitions S whose complements are connected in T.

Tamir [130] discusses *network synthesis games* which include mcst-games.

In [88] v.d. Nouweland et al. study a generalization of mcst-games, *spanning network games*. They show that the class of spanning network games and the class of monotonic games coincide.

Granot and Granot [55] study fixed cost spanning forest problems. In these games the players form a subset of the set of nodes of an undirected graph. They require services which can be provided by facilities that can be constructed at certain nodes of the graph. With each possible facility site a certain cost is associated. There are also costs associated with the use of an edge to deliver the service. In general these games can have an empty core. Granot and Granot show that if the underlying graph is a tree, the game will have a non-empty core.

Aarts [1] studies *chain games*. These are mcst-games that have a minimum cost spanning tree that is a *chain*. A chain is a tree rooted at 0, in which each node, except the last one which is a leaf, has exactly one immediate follower.

S	$c(S)$	S	$c(S)$	S	$c(S)$	S	$c(S)$
$\{1\}$	0	$\{1,2\}$	1	$\{2,4\}$	2	$\{1,3,4\}$	1
$\{2\}$	1	$\{1,3\}$	1	$\{3,4\}$	2	$\{2,3,4\}$	2
$\{3\}$	1	$\{1,4\}$	0	$\{1,2,3\}$	1	$\{1,2,3,4\}$	1
$\{4\}$	1	$\{2,3\}$	1	$\{1,2,4\}$	1	$\emptyset$	0

Table 6.7 An information graph game c.

He gives an expression for the Shapley-value of these games and derives results on their cores.

Kuipers [77] studies *information graph games*. In an information graph game a subset Z of the player set N has a piece of information that is valuable to all the players. Players not in Z can purchase the information from a supplier 0 by paying a fixed price, say 1. Two players who have a communication link between them can, if they decide to cooperate, pass the information from one to the other without any costs. An *information graph* $G_I = (N, E_I)$ describes these communication links. That is, $\{i, j\} \in E_I$ if and only if there is a communication link between i and j. For each coalition $S \subset N$ we consider the subgraph G_I^S of G_I that contains only the nodes in S. This gives us a partition of S into connected components. Within these components the information can be dispersed without any cost as long as there is one player in the component that has the information. If no player in the component has the information then it has to be purchased from the supplier at cost 1 first, and then dispersed among all members of the component. So the cost $c(S)$ of S in the information graph game is equal to the number of components of S that have no customer in Z. An information graph game is a mcst-game. Define the mcst-problem (G, k) on the complete graph $G = (N_0, E)$ by setting

$$
k_{ij} = \begin{cases} 0 & \text{if } \{i,j\} \in E_I \\ 1 & \text{if } \{i,j\} \notin E_I \\ 0 & \text{if } i = 0 \text{ and } j \in Z \\ 1 & \text{if } i = 0 \text{ and } j \notin Z. \end{cases}
$$

The mcst-game given by G has the same characteristic function as the information graph game given by G_I.

Example Let $N = \{1,2,3,4\}$, $Z = \{1\}$, $E_I = \{\{1,4\},\{2,3\}\}$. The information graph game arising from this is given in table 6.7. In table 6.8 the costs for the mcst-game corresponding to the information graph game c are given. The reader van verify easily that the mcst-game given by these costs is indeed the same as the information graph game.

	0	1	2	3	4
0	0	0	1	1	1
1	0	0	1	1	0
2	1	1	0	0	1
3	1	1	0	0	1
4	1	0	1	1	0

Table 6.8 The costs for the mcst-game corresponding to the information graph game in table 6.7.

Kuipers gives conditions under which an information graph game is concave. He shows how to construct for each information graph game a concave information graph game with the same core. In this way he arrives at a characterization of the extreme points of an information graph game. He also provides an efficient way to compute the nucleolus of an information graph game.

Feltkamp et al. [47, 48] consider *minimum cost spanning extension problems* (mcse-problems) and the corresponding mcse-games. In a mcse-game part of a network is already present. Each coalition can use the edges that are initially present and has to construct other edges in order to connect all its members to the source 0.

Granot et al. [60] give a characterization of the nucleolus of a *standard tree game*. They also discuss an algorithm to compute it.

LOCATION GAMES

7.1   INTRODUCTION

Four communities, Amesville, Bethany, Charlestown and Drenton have obtained permission from the county, to which all four of them belong, to build hospitals to service their residents. The county is paying for the costs of building the hospitals under certain conditions. It will not pay for a single community to build a hospital. The communities will have to share hospitals. The county allows any group of two communities to build one hospital, any group of three communities may build two hospitals, and the four communities together may build three hospitals. If a community is not serviced by a hospital it has to fly its patients to a hospital further away at an extremely high cost. The four communities are connected by existing roads. The layout of the four communities is given in figure 7.1. The distances are given in ten miles, i.e., Amesville and Charlestown are connected by a 30 miles long road. The communities can build a hospital anywhere along a road, including at the site of a community.

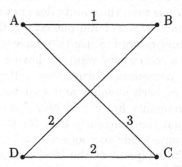

Figure 7.1   The layout of the four communities.

S	$c_p(S)$	S	$c_p(S)$	S	$c_p(S)$	S	$c_p(S)$
$\{1\}$	1	$\{1,2\}$	$\frac{1}{2}$	$\{2,4\}$	1	$\{1,3,4\}$	1
$\{2\}$	1	$\{1,3\}$	$1\frac{1}{2}$	$\{3,4\}$	1	$\{2,3,4\}$	1
$\{3\}$	1	$\{1,4\}$	$1\frac{1}{2}$	$\{1,2,3\}$	$\frac{1}{2}$	$\{1,2,3,4\}$	$\frac{1}{2}$
$\{4\}$	1	$\{2,3\}$	2	$\{1,2,4\}$	$\frac{1}{2}$	$\emptyset$	0

Table 7.1 The p-center game c_p for the four communities in figure 7.1.

Since it can be a matter of life or death, each community wants to minimize the distance to a hospital to which it has access. If a coalition of communities works together they want to minimize the maximum distance between each community in the coalition and the nearest hospital. So costs are considered to be proportional to distances and distances are measured along the existing roads. Suppose that each community has a cost of 1 if no hospital is built. So if they do not cooperate at all their total cost will be 4. If Amesville and Bethany decide to cooperate the best thing to do for them is to build a hospital exactly at the midpoint of the road connecting them. This will make their cost equal to $\frac{1}{2}$. If Amesville and Drenton decide to cooperate they should build a hospital on the road connecting Bethany and Drenton, at the point that is at a distance of $\frac{1}{2}$ from Bethany and $1\frac{1}{2}$ from Drenton. This will make their costs equal to $1\frac{1}{2}$. Continuing in this way we obtain the so-called *p-center game* given in table 7.1. Here Amesville=1, Bethany=2, Charlestown=3, and Drenton=4. Also, p refers to the fact that each coalition S is allowed to build p_S hospitals. So here we have $p_S = |S| - 1$. It is easy to see that this game has a non-empty core.

The same four communities are involved in a project to build telephone switching centers. Again, the county will pay for building the centers, but the communities have to take care of the wirings needed to complete all connections. Similarly as in the previous case, the county does not subsidize a single community, it allows two communities to build one center, three to build two centers, and all four to build three centers. Since they have to take care of all the costs involved in wiring, each community wants to have a center as near by as possible. If a coalition of communities forms, they will want to minimize the sum of the distances between each member and a center nearest to that member. Suppose that each community has a cost of 2 if no center is built. This can be seen as the costs that they currently have for their telephone connections. If Amesville and Bethany decide to cooperate, a best thing for them to do is to build a center at the midpoint of the road connecting them, making their total cost equal to 1. If Amesville, Charlestown, and Drenton cooperate, a best thing for them to do is to build one center at Amesville, and one center at the

S	$m_p(S)$	S	$m_p(S)$	S	$m_p(S)$	S	$m_p(S)$
{1}	2	{1,2}	1	{2,4}	2	{1,3,4}	2
{2}	2	{1,3}	3	{3,4}	2	{2,3,4}	2
{3}	2	{1,4}	3	{1,2,3}	1	{1,2,3,4}	1
{4}	2	{2,3}	4	{1,2,4}	1	$\emptyset$	0

Table 7.2 The *p*-median game m_p for the communities in figure 7.1.

midpoint of the road connecting Charlestown and Drenton, making their total cost equal to 2. Continuing in this way, we arrive at the so-called *p-median game* given in table 7.2. Again, it is easy to see that this game has a non-empty core.

In the following sections we will study the two classes of games introduced here in more detail. In section 7.2 we will introduce them formally and study their balancedness. In section 7.3 we will consider conditions that will guarantee that they are 1-concave or semiconcave. In sections 7.4, 7.5, and 7.6 we will restrict ourselves to location games on trees. There will also be setup costs involved in the games that we consider in these sections.

7.2 LOCATION GAMES WITH EQUAL SETUP COSTS

Let $G = (N, E)$ be a graph, where N, the set of nodes, is the same as the set of players in the game. In the following we assume that G is connected. Every edge $e \in E$ has a positive length l_e. The *distance* $d(x, y)$ between two points x, y anywhere on the edges of the graph is defined as the length of a shortest path from x to y. The length of a path is the sum of the lengths of the edges and parts of edges that belong to the path. Let A be a finite subset of points anywhere on the edges of G and let $i \in N$. The distance $d(i, A)$ between i and A is defined by

$$d(i, A) := \min_{x \in A} d(i, x).$$

The players can construct service facilities at any point on the graph, that is, at any point along an edge of the graph and not only at the nodes of the graph. The cost of a player $i \in N$ is a linear function of the distance between i and a facility that is closest to i. For each $i \in N$ a "weight" w_i is given such that if this distance is d_i, the cost for i is $w_i d_i$. In the context of the examples of the introduction one can think of w_i as depending on the size of the population of community i. We denote the number of facilities that coalition S is allowed to build by p_S and we assume that $p_S < |S|$. Each player i has a certain (high) cost $L(i)$ associated with not having access to any facility. We study two classes of games arising from such a situation which were first introduced in [18]. In the first each coalition wants to minimize the maximum cost of its members, while in the second each coalition wants to minimize the sum of the costs of its members.

Definition 7.2.1 *The characteristic function of a p-center game c_p is given by*

$$c_p(S) := \begin{cases} L(S) & \text{if } p_S = 0 \\ min_{A:|A|=p_S} max_{i \in S} w_i d(i, A) & \text{if } p_S > 0. \end{cases}$$

Here $L(S) := max_{i \in S} L(i)$.

Definition 7.2.2 *The characteristic function of a p-median game m_p is given by*

$$m_p(S) := \begin{cases} \overline{L(S)} & \text{if } p_S = 0 \\ min_{A:|A|=p_S} \sum_{i \in S} w_i d(i, A) & \text{if } p_S > 0. \end{cases}$$

Here $\overline{L(S)} = \sum_{i \in S} L(i)$.

Depending on the situation under consideration a p-center game or a p-median game will be more appropriate.

In the following we will assume that $L(i) \geq c_p(N)$ and $L(i) \geq m_p(N)$ for all $i \in N$. This reflects the fact that it is very important for each player to have access to a facility.

In a p-center game each coalition S with $p_S > 0$ has to solve a p-center problem, whereas in a p-median game each coalition S with $p_S > 0$ has to solve a p-median game. Kariv and Hakimi have shown in [71] and [72] that the problems of finding p-centers and p-medians for $p > 1$ in a given graph, are $\mathcal{NP}$-hard. Nonetheless, it is possible to show that the corresponding games have a non-empty core under certain conditions.

Let us consider the cases where p_S is the same for all coalitions S of equal size. Specifically, let $1 \leq k < |N|$ and let $p_S = (|S| - k)_+$. Then both the p-center game as well as the p-median game are balanced.

Proposition 7.2.3 *Let $1 \leq k \leq |N|$ and let $p_S = (|S| - k)_+$. Then $C(c_p) \neq \emptyset$ and $C(m_p) \neq \emptyset$.*

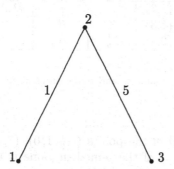

Figure 7.2 A graph leading to a p-center and a p-median game.

Proof. Let S be a coalition with $|S| > k$. Let A_S be such that $|A_S| = p_S$ and

$$c_p(S) = \max_{i \in S} w_i d(i, A_S).$$

Arrange the p_N facilities that N may construct as follows. Put p_S of them at the p_S locations given by A_S. Then there are

$$p_N - p_S = |N| - k - (|S| - k) = |N| - |S| = |N \setminus S|$$

facilities left for N to locate. Let A_N^S be defined by $A_N^S := A_S \cup N \setminus S$. Then $|A_N^S| = p_N$ and

$$
\begin{aligned}
c_p(N) &= \min_{A: |A| = p_N} \max_{i \in N} w_i d(i, A) \\
&\leq \max_{i \in N} w_i d(i, A_N^S) \\
&= c_p(S)
\end{aligned}
$$

It follows that $c_p(N) \leq c_P(S)$ for all $S \in 2^N \setminus \{\emptyset\}$, so any $x \in R^n$ with $x(N) = c_p(N)$ and $x_i \geq 0$ is an element of $C(c_p)$.
The proof that $C(m_p) \neq \emptyset$ is similar to the proof given above. $\qquad\square$

Remark Note that the proof of proposition 7.2.3 holds whenever $p_N \geq p_S + |N \setminus S|$ for all $S \subset N$ with $p_S > 0$.

Example Let $N = \{1, 2, 3\}$ and let the graph G be given in figure 7.2. Let $p_S = (|S| - 1)_+$ and let $L(i) = 1$ for all $i \in N$. The p-center game c_p and the p-median game m_p are given in table 7.3. The core $C(c_p)$ of the p-center

S	$c_p(S)$	$m_p(S)$	S	$c_p(S)$	$m_p(S)$	S	$c_p(S)$	$m_p(S)$
$\{1\}$	1	1	$\{1,2\}$	$\frac{1}{2}$	1	$\{1,2,3\}$	$\frac{1}{2}$	1
$\{2\}$	1	1	$\{1,3\}$	3	6	$\emptyset$	0	0
$\{3\}$	1	1	$\{2,3\}$	$2\frac{1}{2}$	5			

Table 7.3 The p-center game c_p and p-median game m_p arising from figure 7.2.

game c_p is the convex hull of the points $(-\frac{1}{2}, 1, 0)$, $(1, -\frac{1}{2}, 0)$, $(-\frac{3}{2}, 1, 1)$, and $(1, -\frac{3}{2}, 1)$. The core $C(m_p)$ of the p-median game is the convex hull of the points $(0, 1, 0)$, $(1, 0, 0)$, $(-1, 1, 1)$, and $(1, -1, 1)$.

In general the computation of the nucleolus and the τ-value for these games is cumbersome. In the following section we will consider two classes of location games for which this is more manageable.

7.3   1-CONCAVE AND SEMICONCAVE LOCATION GAMES

Recall that a game c is called 1-concave if

$$0 \geq g^c(N) \geq g^c(S) \text{ for all } S \in 2^N \setminus \{\emptyset\}.$$

The results stated in theorems 1.9.6 and 1.9.7 for 1-convex games also hold for 1-concave games. So the core of a 1-concave game c is the convex hull of the vectors $M^c - g^c(N)f^i$, where f^i is the vector with 1 in the i-th place and 0 everywhere else. Also, the nucleolus and τ-value of c coincide and

$$\nu_i(c) = \tau_i(c) = M_i^c - \frac{1}{n}g^c(N).$$

In the following theorem a class of 1-concave location games is identified.

Theorem 7.3.1 Let $p_S = (|S| - (n-2))_+$. For the p-center game c_p we assume that $L(i) \geq c_p(N) - M^{c_p}(N \setminus \{i\})$ for all $i \in N$, and for the p-median game m_p we assume that $L(i) \geq m_p(N) - M^{m_p}(N \setminus \{i\})$ for all $i \in N$. Then the games c_p and m_p are 1-concave.

Proof. Let c_p be a p-center game with $p_S = (|S| - (n-2))_+$ and with $L(i)$ satisfying the condition mentioned above. Then for all $i \in N$ we have

$$
\begin{aligned}
g^{c_p}(N) &= g^{c_p}(N \setminus \{i\}) \\
&= M^{c_p}(N) - c_p(N) \\
&= (n-1)c_p(N) - \sum_{j \in N} c_p(N \setminus \{j\}),
\end{aligned}
$$

and for S with $|S| \leq n-2$ and $i \in S$ we have

$$
\begin{aligned}
g^{c_p}(S) &= M^{c_p}(S) - L(S) \\
&\leq M^{c_p}(S) + M^{c_p}(N \setminus \{i\}) - c_p(N) \\
&\leq M^{c_p}(i) + M^{c_p}(N \setminus \{i\}) - c_p(N) \\
&= M^{c_p}(N) - c_p(N) \\
&= g^{c_p}(N).
\end{aligned}
$$

It follows that $0 \geq g^{c_p}(N) \geq g^{c_p}(S)$ for all $S \subset N$ and hence the game c_p is 1-concave. The proof that the p-median location game is 1-concave proceeds similarly. $\square$

Example The games from table 7.3 are not 1-concave because the $L(i)$'s do not satisfy the condition stated in theorem 7.3.1. If we set $L(1) = 3$, $L(2) = 2\frac{1}{2}$, and $L(3) = 5$ the new p-center game is 1-concave. The core of the new p-center game is the convex hull of the points $(-2, 2\frac{1}{2}, 0)$, $(3, -2\frac{1}{2}, 0)$, and $(-2, -2\frac{1}{2}, 5)$. The nucleolus and the τ-value are equal to $(-\frac{1}{3}, -\frac{5}{6}, \frac{5}{3})$.
If we set $L(1) = 6$, $L(2) = 5$, and $L(3) = 10$ then the new p-median game is 1-concave. The core of the new p-median game is the convex hull of $(-4, 5, 0)$, $(6, -5, 0)$, and $(-4, -5, 10)$. The nucleolus and the τ-value are equal to $(-\frac{2}{3}, -\frac{5}{3}, \frac{10}{3})$.

Example Let $N = \{1, 2, 3, 4\}$ and let the graph G be given in figure 7.3. Let $p_S = (|S| - 1)_+$, $w_i = 1$ for all $i \in N$ and $L(1) = 3$, $L(2) = 2$, $L(3) = 4$, $L(4) = 4$. The p-center game c_p generated by these is given in table 7.4. The $L(i)$'s satisfy the condition in theorem 7.3.1, but the other condition in the theorem is not satisfied and c_p is not 1-concave. ($g^{c_p}(\{3, 4\}) = -3 > -4 = g^{c_p}(N)$.)

Another class of games for which it is relatively easy to compute the τ-value is the class of semiconcave games. Recall that a game c is semiconcave if c is subadditive and $g^c(i) \geq g^c(S)$ for all $i \in N$ and $S \in 2^N$ with $i \in S$. The result stated in theorem 1.9.3 for semiconvex games also holds for semiconcave games,

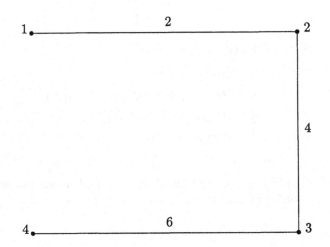

Figure 7.3   A graph leading to a p-center game that is not 1-concave.

S	$c_p(S)$	S	$c_p(S)$	S	$c_p(S)$	S	$c_p(S)$
{1}	3	{1,2}	1	{2,4}	5	{1,3,4}	3
{2}	2	{1,3}	3	{3,4}	3	{2,3,4}	2
{3}	4	{1,4}	6	{1,2,3}	1	{1,2,3,4}	1
{4}	4	{2,3}	2	{1,2,4}	1	$\emptyset$	0

Table 7.4   The p-center game generated by figure 7.3.

so the τ-value $\tau(c)$ of a semiconcave game c is equal to $\lambda\underline{c} + (1 - \lambda)M^c$ where $\underline{c} = (c(1), c(2), \ldots, c(n))$ and $0 \leq \lambda \leq 1$ is such that efficiency is satisfied. In the following theorem we identify a class of semiconcave location games.

Theorem 7.3.2 *Let $p_S = (|S| - (n-1))_+$. We assume that $L(i) \geq c_p(N)$ for all $i \in N$ and that $L(i) \geq m_p(N)$ for all $i \in N$. Then the p-center game and the p-median game are both semiconcave.*

Proof. Let c_p be a p-center game with $p_S = (|S| - (n-1))_+$. Then $c_p(S) = L(S)$ for all coalitions S with $|S| \leq n - 1$. So for all $S \neq N$ we have

$$
\begin{aligned}
g^{c_p}(S) &= M^{c_p}(S) - c_p(S) \\
&= |S|c_p(N) - \sum_{j \in S} L(N \setminus \{j\}) - L(S).
\end{aligned}
$$

It follows that $g^{c_p}(i) \geq g^{c_p}(S)$ for $i \in S$. Also,

$$
\begin{aligned}
g^{c_p}(N) &= M^{c_p}(N) - c_p(N) \\
&= (n-1)c_p(N) - \sum_{j \in N} L(N \setminus \{j\}).
\end{aligned}
$$

Since $L(N \setminus \{j\}) \geq L(i)$ for all $j \in N \setminus \{i\}$ we obtain

$$
\begin{aligned}
g^{c_p}(i) &= c_p(N) - L(N \setminus \{i\}) - L(i) \\
&\geq \sum_{j \in N \setminus \{i\}} (c_p(N) - L(N \setminus \{j\})) - L(N \setminus \{i\}) \\
&= (n-1)c_p(N) - \sum_{j \in N} L(N \setminus \{j\}) \\
&= g^{c_p}(N).
\end{aligned}
$$

Since it is clear that c_p is subadditive, it follows that it is a semiconcave game. The proof that the p-median location game is semiconcave runs similarly. $\square$

If all the $L(i)$'s are the same, the expression for the τ-value of the p-center game and the p-median game becomes very simple.

Corollary 7.3.3 *Let c_p be a p-center game and m_p a p-median game which satisfy the conditions in theorem 7.3.2. Furthermore, let $L(i) = L(j)$ for all $i, j \in N$. Then*

$$
\tau(c_p) = \frac{1}{n}(c_p(N), \ldots, c_p(N)) \text{ and } \tau(m_p) = \frac{1}{n}(m_p(N), \ldots, m_p(N)).
$$

Depts.	Ethno-music.	French	Geography	Health	Interdisc. S.
Costs	2	1	4	2	2

Table 7.5   The setup costs for the computer labs.

Proof. The result follows from theorems 1.9.3 and 7.3.2 and the fact that $c_p(i) = c_p(j)$ $(m_p(i) = m_p(j))$ and $M_i^{c_p} = M_j^{c_p}$ $(M_i^{m_p} = M_j^{m_p})$ for all $i, j \in N$.
□

In the following sections we will restrict our attention to location games on trees with different setup costs.

7.4   SIMPLE PLANT LOCATION GAMES

Five academic departments, the Department of Ethno-musicology, the French Department, the Geography Department, the Health Department, and the Interdisciplinary Studies Department want to build new computer labs that can be used by their faculty and students. The university allows a department to erect labs in any space already assigned to one of the departments. However, the university will not come up with any extra money for the construction of these labs. The departments will have to use the money from their own budget to do this. Depending on where a lab is constructed their are setup costs involved. These setup costs (in $100,000) are given in table 7.5. The departments expect that they will be able to use any lab constructed now for the next ten years without any major additional costs. The layout of the departments with the corridors, staircases and elevators connecting them can be described by a tree. This tree is given in figure 7.4. The numbers along the edges are the estimated costs for ten years (in $1000) per person, involved in going from a department at one endpoint of an edge to the department at the other endpoint. These costs can arise, for example, from time lost in travelling. The departments want to minimize the sum of the setup costs and the costs incurred in travelling. The travel costs have to be weighted to reflect that different numbers of people may be travelling for different departments. Here we assume that all the departments are of approximately the same size, each one having a total number of 100 faculty and students who will be using the labs, so the weights are equal to 100. The departments can decrease their total costs by cooperating. For example, if the Departments of Ethno-musicology and French decide to work together, the best thing for them to do, is to build

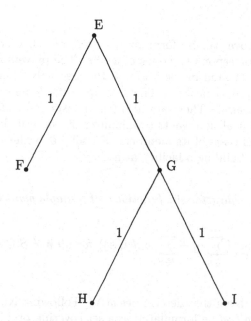

Figure 7.4 The layout of the five departments.

S	$q(S)$	S	$q(S)$	S	$q(S)$	S	$q(S)$
{1}	2	{1,5}	4	{1,2,4}	4	{3,4,5}	5
{2}	1	{2,3}	3	{1,2,5}	4	{1,2,3,4}	5
{3}	3	{2,4}	3	{1,3,4}	5	{1,2,3,5}	5
{4}	2	{2,5}	3	{1,3,5}	5	{1,2,4,5}	6
{5}	2	{3,4}	3	{1,4,5}	6	{1,3,4,5}	7
{1,2}	2	{3,5}	3	{2,3,4}	4	{2,3,4,5}	6
{1,3}	3	{4,5}	4	{2,3,5}	4	{1,2,3,4,5}	7
{1,4}	4	{1,2,3}	4	{2,4,5}	5	∅	0

Table 7.6 The simple plant location game q arising from figure 7.4.

one lab at the French Department. This will make their total costs $200,000. If the French, Geography, and Interdisciplinary Studies Departments decide to cooperate they should build one lab at the French Department, and one lab at the Interdisciplinary Studies Department. This will make their total costs $400,000. The cooperative cost game q arising from this situation is called a *simple plant location game* and is given in table 7.6. The core of this game is not empty. Here Ethno-musicology=1, French=2, Geography=3, Health=4, and Interdisciplinary Studies=5, and the numbers represent costs in $100,000. A core element is (1,1,1,2,2).

The example given above can be formalized as follows. Let N be a set of players. The players correspond to nodes of a tree. The players need to build facilities which can be located in the nodes of the tree only. There are setup costs involved in building a facility. These setup costs depend on the node where the facility is located. There are also travel costs associated with each edge of the tree. Each coalition wants to minimize the sum of the setup costs and the weighted travel costs of its members. For all $j \in N$, let o_j denote the setup costs involved in building a facility at node j.

Definition 7.4.1 *The characteristic function q of a simple plant location game is given by*

$$q(S) := \min_{\emptyset \neq A \subset N} \left(\sum_{j \in A} o_j + \sum_{i \in S} w_i d(i, A) \right) \text{ for all } \emptyset \neq S \subset N.$$

Let us consider the problem that determines $q(N)$. Following Kolen in [74] we show that this problem can be formulated as a set covering problem. For each node $i \in N$ let $0 = r_{i1} \leq r_{i2} \leq \ldots \leq r_{in}$ be the ordered sequence of distances between node i and all the nodes, including i. Define r_{in+1} to be a number that is much larger then the sum of all setup costs and travel costs that are involved. Define the $n^2 \times n$-matrix $H = [h_{i_k j}]$ by

$$h_{i_k j} = \begin{cases} 1 & \text{if } d(i, j) \leq r_{ik} \text{ for } i, j, k \in N \\ 0 & \text{otherwise} \end{cases}$$

The set covering formulation of the simple plant location problem for the grand coalition N is given by

$$
\begin{array}{lll}
\min & \sum_{j=1}^n o_j x_j + \sum_{i=1}^n \sum_{k=1}^n w_i(r_{ik+1} - r_{ik})z_{ik} & \\
\text{s.t.} & \sum_{j=1}^n h_{i_k j} x_j + z_{ik} \geq 1 & \text{for } i, k \in N \\
& z_{ik} \in \{0, 1\} & \text{for } i, k \in N \\
& x_j \in \{0, 1\} & \text{for } j \in N.
\end{array}
\tag{7.1}
$$

Here $x_j = 1$ if and only if a facility is built in node j, and $z_{ik} = 1$ if and only if there is no facility within distance r_{ik} of node i. The number of facilities that are within distance r_{ik} of node i are given by $\sum_{j=1}^n h_{i_k j} x_j$.
Let $d_{ik} = w_i(r_{ik+1} - r_{ik})$. Then we obtain the following formulation for problem 7.1.

$$
\begin{array}{lll}
\min & \sum_{j=1}^n o_j x_j + \sum_{i=1}^n \sum_{k=1}^n d_{ik} z_{ik} & \\
\text{s.t.} & \sum_{j=1}^n h_{i_k j} x_j + z_{ik} \geq 1 & \text{for } i, k \in N \\
& z_{ik} \in \{0, 1\} & \text{for } i, k \in N \\
& x_j \in \{0, 1\} & \text{for } j \in N.
\end{array}
\tag{7.2}
$$

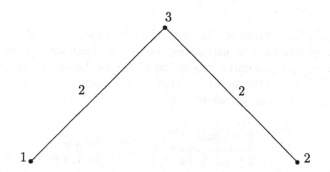

Figure 7.5 A tree leading to a simple plant location game.

For $S \subset N$ the set-covering formulation for $q(S)$ is

$$
\begin{array}{lll}
\min & \sum_{j=1}^{n} o_j x_j + \sum_{i \in S} \sum_{k=1}^{n} d_{ik} z_{ik} & \\
\text{s.t.} & \sum_{j=1}^{h} h_{i_k} x_j x_j + z_{ik} \geq 1 & \text{for } i \in S, k \in N \\
& z_{ik} \in \{0, 1\} & \text{for } i \in S, k \in N \\
& x_j \in \{0, 1\} & \text{for } j \in N.
\end{array} \qquad (7.3)
$$

Example let $N = \{1, 2, 3\}$ and let the graph G be as given in figure 7.5. For this game we have

$$
\begin{array}{lll}
r_{11} = 0 & r_{12} = 2 & r_{13} = 3 \\
r_{21} = 0 & r_{22} = 1 & r_{23} = 3 \\
r_{31} = 0 & r_{32} = 1 & r_{33} = 2
\end{array}
$$

and

$$
H = \begin{pmatrix}
1 & 0 & 0 \\
0 & 1 & 0 \\
0 & 0 & 1 \\
1 & 0 & 1 \\
0 & 1 & 1 \\
0 & 1 & 1 \\
1 & 1 & 1 \\
1 & 1 & 1 \\
1 & 1 & 1
\end{pmatrix}
$$

Here the first three rows of H correspond to $k = 1$, the next three rows to $k = 2$ and the last three rows to $k = 3$.

In general the set-covering problem is $\mathcal{NP}$-hard. In [74] Kolen showed that the set-covering problem corresponding to the simple plant location problem can

be solved in polynomial time, due to the fact that the matrix H can be brought
in standard greedy form by permuting the rows and columns. This also implies
that the linear programming relaxation of the problem will yield an integer
solution. It follows that $q(N)$ is equal to the value of the dual problem of
problem 7.2 which is given below.

$$
\begin{array}{lll}
\max & \sum_{i=1}^{n} \sum_{k=1}^{n} y_{ik} & \\
\text{s.t.} & \sum_{i=1}^{n} \sum_{k=1}^{n} y_{ik} h_{i_k j} \leq o_j & \text{for } j \in N \\
& 0 \leq y_{ik} \leq d_{ik} & \text{for } i, k \in N.
\end{array}
\tag{7.4}
$$

This result can be used to obtain a core element for the simple plant location
game.

Theorem 7.4.2 *Let q be a simple plant location game. Then $C(q) \neq \emptyset$.*

Proof. Let $\hat{y}$ be a solution for problem 7.2. Define $u \in R^n$ by

$$
u_i := \sum_{k=1}^{n} \hat{y}_{ik} \text{ for all } i \in N.
$$

Then

$$
u(N) \quad = \quad q(N) \text{ and } u(S) \leq q(S) \text{ for all } S \subset N.
$$

Here the inequality follows because $\hat{y}$ is also feasible for the dual problem of the
linear programming relaxation of the set-covering problem 7.3 that determines
$q(S)$. So $u \in C(q)$.

7.5 MEDIAN GAMES WITH BUDGET CONSTRAINTS

Let us consider the example of the five departments again. We will make the
following alteration. Each department has a certain budget set aside for the
construction of the computer labs. A department does not care how much
the setup costs are, as long as they do not exceed its budget. The budgets of
the five departments (in $100,000) are given in table 7.7. If the Department
of Ethno-musicology does not cooperate with any other department then the
only place at which it can build a computer lab is at the site corresponding
to the French Department. It will have to do this, or use some other, already

Depts.	Ethno-music.	French	Geography	Health	Interdisc. S.
Budget	1	2	1	1	2

Table 7.7   The budgets of the five departments.

Depts.	Ethno-music.	French	Geography	Health	Interdisc. S.
$L(i)$	5	3	6	4	4

Table 7.8   The costs of each department if it has to use an already existing lab.

S	$m(S)$	S	$m(S)$	S	$m(S)$	S	$m(S)$
{1}	1	{1,5}	2	{1,2,4}	1	{3,4,5}	1
{2}	0	{2,3}	2	{1,2,5}	0	{1,2,3,4}	1
{3}	1	{2,4}	0	{1,3,4}	3	{1,2,3,5}	1
{4}	2	{2,5}	0	{1,3,5}	1	{1,2,4,5}	1
{5}	0	{3,4}	1	{1,4,5}	2	{1,3,4,5,}	3
{1,2}	0	{3,5}	1	{2,3,4}	1	{2,3,4,5}	1
{1,3}	1	{4,5}	2	{2,3,5}	1	{1,2,3,4,5}	1
{1,4}	2	{1,2,3}	1	{2,4,5}	0	∅	0

Table 7.9   A median game with budget constraints.

existing, lab which will involve very high travel costs, say \$5000 per person per 10 years. In table 7.8 the travel costs (in \$1000) per person per 10 years, of each department if it has to use an already existing lab are given. The cooperative cost game m arising from this situation is called a *median game with budget constraints* and is given in table 7.9. The point $(0,0,0,1,0)$ is a core element for this game.

In general, median games with budget constraints correspond to situations in which the number of facilities that a coalition is allowed to build is not prescribed but is limited by the budget of the coalition. Every player i has a budget b_i and the budget of coalition S is $b(S) = \sum_{i \in S} b_i$.

Definition 7.5.1 *The characteristic function m of a median game with budget constraints is given by*

$$m(S) := \begin{cases} min_{A \subset N, o(A) \leq b(S)} \sum_{i \in S} w_i d(i, A) & \text{if } b(S) \geq min_{j \in N} o_j \\ \overline{L(S)} & \text{otherwise.} \end{cases}$$

here $\overline{L(S)}$ is as defined in definition 7.2.2.

The median game with budget constraints and the simple plant location game are closely related. In both cases a coalition wants to minimize the sum of the travel costs of its members, but whereas in the simple plant location game the setup costs appear in the objective function as costs to be minimized too, in the median game with budget constraints they appear as constraints. The following 0-1 programming problem is a modification of 7.1 which determines $q(N)$, and its value is equal to $m(N)$.

$$\begin{aligned} \min \quad & \sum_{i=1}^{n} \sum_{k=1}^{n} w_i (r_{ik+1} - r_{ik}) z_{ik} \\ \text{s.t.} \quad & \sum_{j=1}^{n} h_{i_k j} x_j + z_{ik} \geq 1 & \text{for } i, k \in N \\ & \sum_{j=1}^{n} o_j x_j \leq b(N) & \\ & z_{ik} \in \{0,1\} & \text{for } i, k \in N \\ & x_j \in \{0,1\} & \text{for } j \in N. \end{aligned} \qquad (7.5)$$

Because of the extra constraint $\sum_{j=1}^{n} o_j x_j \leq b(N)$ problem 7.5 cannot be treated in the same way as problem 7.1. In fact, unlike the simple plant location game the median game with budget constraints can have an empty core as the following example shows.

Example Let $N = \{1, 2, 3, 4, 5\}$ and let the tree G be as given in figure 7.6. The setup costs o_i, alternative costs $L(i)$, budgets b_i, and weights w_i

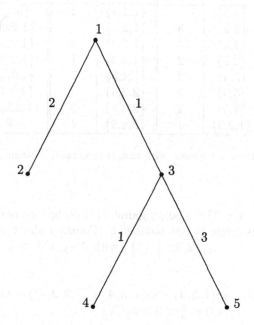

Figure 7.6 A tree leading to a non-balanced median game with budget constraints.

i	o_i	$L(i)$	w_i	b_i
1	1	5	1	1
2	2	5	1	1
3	3	5	1	1
4	3	5	1	1
5	1	5	1	1

Table 7.10 The setup costs, alternative costs, budgets and weights for a median game with budgets constraints.

S	$m(S)$	S	$m(S)$	S	$m(S)$	S	$m(S)$
{1}	0	{1,5}	0	{1,2,4}	2	{3,4,5}	4
{2}	2	{2,3}	3	{1,2,5}	2	{1,2,3,4}	3
{3}	1	{2,4}	4	{1,3,4}	2	{1,2,3,5}	1
{4}	2	{2,5}	2	{1,3,5}	1	{1,2,4,5}	2
{5}	0	{3,4}	3	{1,4,5}	2	{1,3,4,5}	2
{1,2}	2	{3,5}	1	{2,3,4}	3	{2,3,4,5}	3
{1,3}	1	{4,5}	2	{2,3,5}	3	{1,2,3,4,5}	3
{1,4}	2	{1,2,3}	1	{2,4,5}	4	$\emptyset$	0

Table 7.11   The median game with budget constraints m arising from figure 7.6.

are given in table 7.10. The median game with budget constraints m is given in table 7.11. This game is not balanced. Consider the balanced collection $\{\{1,2,3\},\{1,2,4\},\{1,3,4\},\{2,3,4\},\{5\}\}$ with $\lambda_{\{5\}} = 1$ and all other weights equal to $\frac{1}{3}$. Then

$$\tfrac{1}{3}(m(1,2,3) + m(1,2,4) + m(1,3,4) + m(2,3,4)) + m(5) =$$
$$\tfrac{1}{3} + \tfrac{2}{3} + \tfrac{2}{3} + \tfrac{3}{3} + 0 = \tfrac{8}{3} < 3 = m(N)$$

It follows that $C(m) = \emptyset$.

From every situation that generates a median game with budget constraints, a simple plant location game can also be constructed. If the best way to put the facilities (the way that minimizes the sum of the weighted travel costs and the setup costs) in this simple plant location game yields setup costs that are equal to the budget for each coalition, then the median game with budget constraints will have a non-empty core.

Proposition 7.5.2 *Let m be a median game with budget constraints. Let q be the corresponding simple plant location game. Suppose that for all $\emptyset \neq S \subset N$ we have*

$$q(S) = \sum_{i \in S} w_i d(i, A) + \sum_{j \in A} o_j \text{ implies } \sum_{j \in A} o_j = b(S).$$

Then $C(m) \neq \emptyset$.

Proof. From the condition in the theorem it follows that $m(S) = q(S) - b(S)$ for all $S \subset N$. Since $C(q) \neq \emptyset$ it follows that $C(m) \neq \emptyset$. $\square$

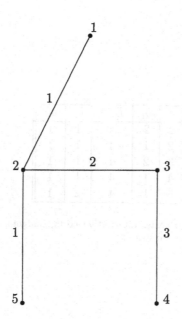

Figure 7.7 A tree leading to a center game with budget constraints.

7.6   CENTER GAMES WITH BUDGET CONSTRAINTS

For the games that we will consider in this section we have the same constraints as we had in section 7.5. The only difference is that now each coalition wants to minimize the maximum weighted distance between any member of the coalition and a facility that is nearest to that member.

Definition 7.6.1 *The characteristic function c of a center game with budget constraints is given by*

$$c(S) := \begin{cases} min_{A \subset N, o(A) \leq b(S)} max_{i \in S} w_i d(i, A) & \text{if } b(S) \geq min_{j \in N} o_j \\ L(S) & \text{otherwise.} \end{cases}$$

Here $L(S)$ *is as defined in definition 7.2.1.*

Example Let $N = \{1, 2, 3, 4, 5\}$ and let G be as in figure 7.7. The setup costs o_i, alternative costs $L(i)$, budgets b_i, and weights w_i are given in table 7.12. The center game with budget constraints c is given in table 7.13. This game is not

i	o_i	$L(i)$	b_i	w_i
1	2	4	1	1
2	1	4	1	1
3	2	4	1	1
4	1	4	1	1
5	1	4	2	1

Table 7.12 The setup costs, alternative costs, budgets, and weights for a center game with budget constraints.

S	$c(S)$	S	$c(S)$	S	$c(S)$	S	$c(S)$
{1}	1	{1,5}	0	{1,2,4}	1	{3,4,5}	0
{2}	0	{2,3}	2	{1,2,5}	0	{1,2,3,4}	1
{3}	2	{2,4}	0	{1,3,4}	2	{1,2,3,5}	1
{4}	0	{2,5}	0	{1,3,5}	1	{1,2,4,5}	0
{5}	0	{3,4}	2	{1,4,5}	0	{1,3,4,5}	1
{1,2}	1	{3,5}	0	{2,3,4}	2	{2,3,4,5}	0
{1,3}	2	{4,5}	0	{2,3,5}	0	{1,2,3,4,5}	1
{1,4}	1	{1,2,3}	1	{2,4,5}	0	$\emptyset$	0

Table 7.13 The center game with budget constraints c arising from figure 7.7.

balanced. Consider the balanced collection $\{\{1,5\}, \{2,4\}, \{3,5\}, \{1,2,3\}, \{4\}\}$ with all weights equal to $\frac{1}{2}$. Then

$$\frac{1}{2}(c(1,5) + c(2,4) + c(3,5) + c(1,2,3) + c(4)) = \frac{1}{2} < 1 = c(N),$$

so c is not balanced.

Although a center game with budget constraints can have an empty core we can define a class of balanced games that are closely related to center games with budget constraints. In addition to the graph G, the setup costs o_i, and the weights w_i, let $r > 0$ be given.

Definition 7.6.2 *The characteristic function of the covering game γ_r is given by*

$$\gamma_r(S) := \begin{array}{ll} min & \sum_{j \in A} o_j \\ s.t. & w_i d(i, A) \le r \quad \text{for all } i \in S \\ & \emptyset \ne A \subset N \end{array}$$

In the center game with budget constraints the players want a facility as close by as possible, with the constraint that the setup costs cannot exceed their budget. In the covering game the players demand that a facility be located not farther away than a distance r from them, and they want to achieve this at the least possible cost. Let the $n \times n$-matrix $H = [h_{ij}]$ be defined by

$$h_{ij} := \begin{cases} 1 & \text{if } w_i d(o, j) \le r \\ 0 & \text{otherwise.} \end{cases}$$

Then $\gamma_r(N)$ is equal to the value of the following 0-1 programming problem.

$$\begin{array}{ll} min & \sum_{j=1}^{n} o_j x_j \\ s.t. & \sum_{j=1}^{h} h_{ij} x_j \ge 1 \quad \text{for } i \in N \\ & x_j \in \{0, 1\} \end{array} \tag{7.6}$$

Kolen [74] showed that the matrix H can be brought in standard greedy form by permuting rows and columns. It follows that $\gamma_r(N)$ is equal to the linear programming relaxation of 7.6. Similarly to the way we did it in section 7.4 for a simple plant location game, we can show that the game γ_r has a non-empty core. In fact, the game γ_r is a special case of a game studied by Tamir in [127]. Let R_S be the smallest r such that $\gamma_r(S) \le b(S)$. Then $c(S) = R_s$. This yields the following result.

i	o_i	b_i	b_i'
1	1	2	1
2	2	1	1
3	3	2	1
4	3	2	1
5	1	2	1

Table 7.14 The setup costs and budgets for the center games of the example.

Proposition 7.6.3 *Let c be a center game with budget constraints. Let $R_S \geq R_N$ for all $S \subset N$ for the corresponding covering problems. Then*

$$(\frac{c(N)}{n}, \ldots, \frac{c(N)}{n}) \in C(c).$$

Proof.

$$\sum_{i \in N} \frac{c(N)}{n} = c(N) \text{ and}$$

$$\sum_{i \in S} \frac{c(N)}{n} = |S|\frac{c(N)}{n} < c(N) \leq c(S)$$

for all $S \subset N$.   □

If there exists an S such that $R_N > R_S$ nothing can be said about the core of the game c. As the following example shows there exists a game satisfying this property with an empty core, and there also exists a game satisfying this property with a non-empty core.

Example Let $N = \{1, 2, 3, 4, 5\}$ and let the graph G be given in figure 7.8. Let $w_i = L(i) = 1$ for all $i \in N$. The setup costs and budgets are given in table 7.14. Let c be the center game with budgets given by the b_i's and let c' be the center game with budgets given by the b_i''s. The games c and c' are given in table 7.15. For both games we see that there is a coalition with costs strictly less than those of the grand coalition. We will show that c has an empty core, whereas c' has a non-empty core.
Consider the balanced collection $\{\{2, 3, 4\}, \{1, 2\}, \{1, 3\}, \{1, 4\}\}$ with weights $\lambda_{\{2,3,4\}} = \frac{2}{3}$, and $\lambda_{\{1,2\}} = \lambda_{\{1,3\}} = \lambda_{\{1,4\}} = \frac{1}{3}$. We see that

$$\frac{2}{3}c(2, 3, 4) + \frac{1}{3}(c(1, 2) + c(1, 3) + c(1, 4)) = \frac{2}{3} < 1 = c(N).$$

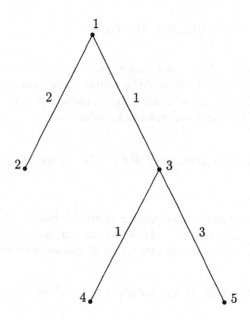

Figure 7.8 The tree for the center games with budget constraints of the example.

S	$c(S)$	$c'(S)$	S	$c(S)$	$c'(S)$	S	$c(S)$	$c'(S)$
{1}	0	0	{2,5}	0	2	{2,3,5}	1	2
{2}	2	2	{3,4}	1	2	{2,4,5}	2	2
{3}	1	1	{3,5}	0	1	{3,4,5}	1	2
{4}	2	2	{4,5}	0	2	{1,2,3,4}	1	2
{5}	0	0	{1,2,3}	1	1	{1,2,3,5}	0	1
{1,2}	0	2	{1,2,4}	1	2	{1,2,4,5}	0	2
{1,3}	0	1	{1,2,5}	0	2	{1,3,4,5}	0	1
{1.4}	0	2	{1,3,4}	1	1	{2,3,4,5}	1	2
{1,5}	0	0	{1,3,5}	0	1	{1,2,3,4,5}	1	2
{2,3}	1	2	{1,4,5}	0	2	∅	0	0
{2,4}	2	2	{2,3,4}	1	2			

Table 7.15 The center games with budget constraints c and c'.

So c is not balanced.

The reader can verify that $(0, 1, 0, 1, 0) \in C(c')$.

This example shows that if the condition in proposition 7.6.3 is not satisfied one cannot say anything about the core of the center game with budget constraints. If $R_S \leq R_n$ for all $S \subset N$, we will show that a game which is closely related to the center game with budget constraints is balanced.

Definition 7.6.4 *Let the game* $< N, \delta(S) >$ *be defined by* $\delta(S) := \gamma_{R_S}(S)$ *for all* $\emptyset \neq S \subset N$.

Note that $\delta(S)$ is equal to the costs that S incurs by building facilities in such a way that each of its members has a facility that is at a distance of at most R_S. Remember that R_S is the best distance that S can achieve given its budget.

Proposition 7.6.5 *Let* $R_S \leq R_N$ *for all* $S \subset N$. *Then* $C(\gamma_{R_N}) \subset C(\delta)$.

Proof. Let $x \in C(\gamma_{R_N})$. Then

$$x(N) = \gamma_{R_N}(N) \quad = \quad \delta(N) \text{ and}$$
$$x(S) \leq \gamma_{R_N}(S) \quad \leq \quad \gamma_{R_S}(S) = \delta(S).$$

$\square$

Since $C(\gamma_{R_N}) \neq \emptyset$ it follows that $C(\delta) \neq \emptyset$.

REFERENCES

[1] Aarts, H. (1994) Minimum cost spanning tree games and set games. Ph.D. Thesis, University of Twente, Enschede, The Netherlands.

[2] Aumann, R.J. and J.H. Drèze (1974). Cooperative games with coalition structures. *International Journal of Game Theory* 3, 217-237.

[3] Aumann, R.J. and M. Maschler (1964). The bargaining set for cooperative games. In: *Advances in Game Theory* (M. Dresher, L.S. Shapley, and A. Tucker eds.), Princeton University Press, Princeton, New Jersey, 443-476.

[4] Aumann, R.J. and M. Maschler (1985). Game theoretic analysis of a bankruptcy problem from the Talmud. *Journal of Economic Theory* 36, 195-213.

[5] Balinski, M.L. and D. Gale (1990). On the core of the assignment game. In: *Functional Analysis, Optimization, and Mathematical Economics* Oxford University Press, New York, 274-289.

[6] Banker, R.D. (1981). Equity considerations in traditional full cost allocation practices: An axiomatic perspective. In: *Joint Cost Allocations* (S. Moriarity, ed.). University of Oklahoma, 110-13

[7] Banzhaf, J.F. III (1965) Weighted voting doesn't work: A mathematical analysis. *Rutgers Law Review* 19, 317-343.

[8] Bird, C.G. (1976). On cost allocation for a spanning tree: A game theoretic approach. *Networks* 6, 335-350.

[9] Bolger, E.M. (1980). A class of power indices for voting games. *International Journal of Game Theory* 9, 217-232.

173

[10] Bolger, E.M. (1982). Characterizing the Banzhaf and Shapley values assuming limited linearity. *International Journal of Game Theory* 11, 1-12.

[11] Brams, S.J. (1975). Game theory and politics. Free Press, New York.

[12] Bondareva, O.N. (1963). Certain applications of the methods of linear programming to the theory of cooperative games. (In Russian) *Problemy Kibernetiki* 10, 119-139.

[13] Claus, A. and D.J. Kleitman (1973). Cost allocation for a spanning tree. *Networks* 3, 289-304.

[14] Coleman, J.S. (1971). Control of collectivities and the power of a collectivity to act. In: *Social Choice* (B. Lieberman, ed.). Gordon and Breach, London, 269-300.

[15] Crawford, V.P. and E.M. Knoer (1981). Job matching with heterogeneous firms and workers. *Econometrica* 49, 437-450.

[16] Curiel, I.J. (1987a). A class of non-normalized power indices for simple games. *Mathematical Social Sciences* 13, 141-152.

[17] Curiel, I.J. (1987b). Combinatorial games. In: *Surveys in game theory and related topics* (H.J.M. Peters and O.J. Vrieze eds.) CWI-tract 39. Centre for Mathematics and Computer Science, Amsterdam, 229-250.

[18] Curiel, I.J. (1990). Location Games. Research Report 90-20, Department of Mathematics, University of Maryland Baltimore County.

[19] Curiel, I.J., J.J.M. Derks and S.H. Tijs (1989). On balanced games and games with committee control. *OR Spektrum* 11, 83-88.

[20] Curiel, I.J., H. Hamers, J.A.M. Potters, and S.H. Tijs (1993). The equal gain splitting rule for sequencing situations and the general nucleolus. Research Memorandum FEW 629 Tilburg University, The Netherlands.

[21] Curiel, I.J., M. Maschler and S.H. Tijs (1987). Bankruptcy games. *Zeitschrift für Operations Research Series A* 31, 143-159.

[22] Curiel, I.J., G. Pederzoli and S.H. Tijs (1988a). Reward allocations in production systems. In: *Lecture Notes in Economics and Mathematical Systems* (H.A. Eiselt and G. Pederzoli eds.). Springer-Verlag, Berlin Heidelberg, 186-199.

[23] Curiel, I.J., G. Pederzoli and S.H. Tijs (1989). Sequencing games. *European Journal of Operational Research* 40, 344-351.

[24] Curiel, I.J., J.A.M. Potters, V. Rajendra Prasad, S.H. Tijs, and B. Veltman (1993). Cooperation in one machine scheduling. *Zeitschrift für Operations Research* 38, 113-129.

[25] Curiel, I.J., J.A.M. Potters, V. Rajendra Prasad, S.H. Tijs, and B. Veltman (1995). Sequencing and cooperation. *Operations Research* 42, 566-568.

[26] Curiel, I.J. and S.H. Tijs (1986). Assignment games and permutation games. *Methods of Operations Research* 54, 323-334.

[27] Curiel, I.J. and S.H. Tijs (1991). Minimarg and maximarg operators. *Journal of Optimization Theory and Applications* 71, 277-287.

[28] Davis, M. and M. Maschler (1965). The kernel of a cooperative game. *Naval Research Logistics Quarterly* 12, 223-259.

[29] Deegan, J. and E.W. Packel (1978). A new index of power for simple n-person games. *International Journal of Game Theory* 7, 113-123.

[30] Deegan, J. and E.W. Packel (1983). To the (minimal winning) victors go the (equally divided) spoils: A new power index for simple n-person games. In: *Political and Related Models* (S.J. Brams, W.F. Lucas and P.D. Straffin Jr. eds.). Springer-Verlag, New York, 239-255.

[31] Derks, J.J.M. (1987). Decomposition of games with non-empty core into veto-controlled simple games. *OR Spektrum* 9, 81-85.

[32] Derks, J.J.M. and J. Kuipers (1996). On the core and nucleolus of routing games. *International Journal of Game Theory* (to appear).

[33] Dijkstra, E.W. (1959). A note on two problems in connection with graphs. *Numerische Mathematik* 1, 269-271.

[34] Driessen, T.S.H. (1985a). A new axiomatic characterization of the Shapley-value. *Methods of Operations Research* 50, 505-517.

[35] Driessen, T.S.H. (1985b). Properties of 1-convex n-person games. *OR Spektrum* 7, 19-26.

[36] Driessen, T.S.H. (1988) Cooperative games, solutions and applications. Kluwer Academic Publishers, Dordrecht, The Netherlands.

[37] Driessen, T.S.H. and S.H. Tijs (1984). Extensions and modifications of the τ-value for cooperative games. In: *Selected Topics in Operations Research and Mathematical Economics* (G. Hammer and D. Pallaschke eds.). Springer-Verlag, Berlin, 252-261.

[38] Driessen, T.S.H. and S.H. Tijs (1985). The τ-value, the core and semiconvex games. *International Journal of Game Theory* 14, 229-247.

[39] Dubey, P. (1975). On the uniqueness of the Shapley-value. *International Journal of Game Theory* 4, 131-139.

[40] Dubey, P. and L.S. Shapley (1979). Mathematical properties of the Banzhaf power index. *Mathematics of Operations Research* 4, 99-131.

[41] Dubey, P. and L.S. Shapley (1984). Totally balanced games arising from controlled programming problems. *Mathematical Programming* 29, 245-267.

[42] Dubins, L.E. and D.L. Freedman (1981). Machiavelli and the Gale-Shapley algorithm. *American Mathematical Monthly* 88, 485-494.

[43] Edmonds, J. (1967). Optimum branchings. *Journal of Research of the National Bureau of Standards* 71B, 233-240.

[44] Faigle, U. (1989) Cores of games with restricted cooperation. *Zeitschrift für Operations Research* 33, 405-422.

[45] Feltkamp, V., A. v.d. Nouweland, A. Koster, P. Borm, and S.H. Tijs (1993). Linear production with transport of products, resources and technology. *Zeitschrift für Operations Research* 38, 153-162.

[46] Feltkamp, V., S.H. Tijs, and S. Muto (1994a). Bird's tree allocations revisited. CentER Discussion Paper 9435, Tilburg University, Tilburg, The Netherlands.

[47] Feltkamp, V., S.H. Tijs, and S. Muto (1994b). On the irreducible core and the equal remaining obligations rule of minimum cost spanning extension problems. *CentER Discussion Paper* 9410, Tilburg University, Tilburg, The Netherlands.

[48] Feltkamp, V., S.H. Tijs, and S. Muto (1994c). Minimum cost spanning extension problems: the proportional rule and the decentralized rule. *CentER Discussion Paper* 9496, Tilburg University, Tilburg, The Netherlands.

[49] Ford, L.R. Jr. and D.R. Fulkerson (1962). Flows in Networks. Princeton University Press, Princeton.

[50] Gale, D. (1984). Equilibrium in a discrete exchange economy with money. *International Journal of Game Theory* 13, 61-64.

[51] Gale, D. and L.S. Shapley (1962). College admission and the stability of marriage. *American Mathematical Monthly* 69, 9-15.

[52] Gale, D. and M. Sotomayor (1985). Ms. Machiavelli and the stable matching problem. *American Mathematical Monthly* 92, 261-268.

[53] Gillies, D.B. (1953). Some theorems on *n*-person games. Ph.D. Thesis, Princeton University.

[54] Granot, D. (1986). A generalized linear production model: A unifying model. *Mathematical Programming* 34, 212-223.

[55] Granot, D. and F. Granot (1992). Computational complexity of a cost allocation approach to a fixed cost spanning forest problem. *Mathematics of Operations Research* 17, 765-780.

[56] Granot, D. and Granot F. (1992). On some network flow games. *Mathematics of Operations Research* 17, 792-841.

[57] Granot, D. and G. Huberman (1981). Minimum cost spanning tree games. *Mathematical programming* 21, 1-18.

[58] Granot, D. and G. Huberman(1982). The relationship between convex games and minimal cost spanning tree games: A case for permutationally convex games. *Siam Journal of Algebra and Discrete Mathematics* 3, 288-292.

[59] Granot, D. and G. Huberman (1984). On the core and nucleolus of minimum cost spanning tree games. *Mathematical Programming* 29, 323-347.

[60] Granot, D., M. Maschler, G. Owen, and W.R. Zhu (1995). The kernel/nucleolus of a standard tree game. *International Journal of Game Theory* 25, 219-244.

[61] Hamers, H., P. Borm, and S.H. Tijs (1995). On games corresponding to sequencing situations with ready times. *Mathematical Programming* 69 471-483.

[62] Hamers, H., J. Suijs, S.H. Tijs, and P. Borm (1996). The split core for sequencing games. *Games and Economic Behavior* 15, 165-176.

[63] Hart, S. and A. Mas-Colell (1987). Potential, value, and consistency. *Econometrica* 55, 935-962.

[64] Herstein, I.N. and J. Milnor (1953). An axiomatic approach to measurable utility. *Econometrica* 21, 291-297.

[65] Ichiishi, T. (1981). Super-modularity: Applications to convex games and to the greedy algorithm for LP. *Journal of Economic Theory* 25, 283-286.

[66] Kalai, E. and E. Zemel (1982a). Totally balanced games and games of flow. *Mathematics of Operations Research* 7, 476-478.

[67] Kalai, E. and E. Zemel (1982b). Generalized networks problems yielding totally balanced games. *Operations Research* 30, 998-1008.

[68] Kaneko, M. (1982). The central assignment game and the assignment markets. *Journal of Mathematical Economics* 10, 205-232.

[69] Kaneko, M. (1983). Housing markets with indivisibilities. *Journal of Urban Economics* 13, 22-50.

[70] Kaneko, M. and Y. Yamamoto (1986). The existence and computation of competitive equilibria in markets with an indivisible commodity. *Journal of Economic Theory* 38, 118-136.

[71] Kariv, O. and S.L. Hakimi (1979a) An algorithmic approach to network location problems. I: The p-centers. *SIAM Journal of Applied Mathematics* 37, 513-538.

[72] Kariv, O. and S.L. Hakimi (1979b) An algorithmic approach to network location problems. II: The p-medians. *SIAM Journal of Applied Mathematics* 37, 539-560.

[73] Kilgour, D.M. (1974). A Shapley-value for cooperative games with quarelling. In: *Game Theory as a Theory of Conflict Resolution* (A. Rapoport ed.). D. Reidel, Dordrecht, Holland, 193-206.

[74] Kolen, A.J.W. (1986). Tree networks and planar rectilinear location theory. CWI-tract 25, Centre for Mathematics and Computer Science.

[75] Kruskal, J.B. (1956). On the shortest spanning subtree of a graph and the traveling salesman problem. *Proceedings of the American Mathematical Society* 7, 48-50.

[76] Kuipers, J. (1991). A note on the 5-person traveling salesman game. *Zeitschrift für Operations Research* 21, 339-351.

[77] Kuipers, J. (1993). On the core of information graph games. *International Journal of Game Theory* 21, 339-350.

[78] Kuipers, J. (1994). Combinatorial Methods in Cooperative Game Theory. Ph. D. Thesis, Limburg University, Maastricht, The Netherlands.

[79] Lawler, E.L., J.K. Lenstra, A.H.G. Rinnooy Kan, and D.B. Shmoys, eds. (1985). The traveling salesman problem: A guided tour of combinatorial optimization. Wiley, New York.

[80] Lucas, W.F. (1983). Measuring power in weighted voting systems. In: *Political and Related Models* (S.J. Brams, W.F. Lucas and P.D. Straffin Jr. eds.). Springer-Verlag, New York, 181-238.

[81] Maschler, M., B. Peleg, and L.S. Shapley (1972). The kernel and bargaining set for convex games. *International Journal of Game Theory* 1, 73-93.

[82] Megiddo, N. (1978a). Cost allocation for Steiner trees. *Networks* 8, 1-6.

[83] Megiddo, N. (1978b). Computationally complexity of the game theory approach to cost allocation for a tree. *Mathematics of Operations Research*

3, 189-196.

[84] Moulin, H. (1985). The separability axiom and equal-sharing methods. *Journal of Economic Theory* 36, 120-148.

[85] Myerson, R.B. (1977). Graphs and cooperation in games. *Mathematics of Operations Research* 2, 225-229.

[86] Neumann, J. von and O. Morgenstern (1944). Theory of games and economic behavior. Princeton University Press, Princeton.

[87] O'Neill, B. (1982). A problem of rights arbitration from the Talmud. *Mathematical Social Sciences* 2, 345-371.

[88] Nouweland v.d. A., S.H. Tijs, and M. Maschler (1993). Monotonic games are spanning network games. *International Journal of Game Theory* 21, 419-427.

[89] Owen, G. (1972). Multilinear extensions of games. *Management Science* 18, 64-79.

[90] Owen, G. (1975a). Multilinear extensions and the Banzhaf value. *Naval Research Logistics Quarterly* 22, 741-750.

[91] Owen, G. (1975b). The core of linear production games. *Mathematical Programming* 9, 358-370.

[92] Owen, G. (1977). Values of games with a priori unions. In: *Mathematical Economics and Game Theory* (R. Henn and O. Moeschlin eds.). Berlin, 76-88.

[93] Owen, G. (1981). Modification of the Banzhaf-Coleman index for games with a priori unions. In: *Power, Voting, and Voting Power* (M.J. Holler ed.). Physica-Verlag, Würzburg.

[94] Owen, G. (1982). Game theory (second edition). Academic Press, Orlando.

[95] Owen, G. (1986). Values of graph restricted games. *SIAM Journal on Algebraic an Discrete Methods* 7, 210-220.

[96] Packel, E.W. and J. Deegan (1980). An axiomated family of power indices for simple *n*-person games. *Public Choice* 35, 229-239.

[97] Peleg, B. (1985). On the reduced game property and its converse. *International Journal of Game Theory* 15, 187-200.

[98] Potters, J.A.M. (1989). A class of traveling salesman games. *Methods of Operations Research* 59, 263-276.

[99] Potters, J.A.M., I.J. Curiel, and S.H. Tijs (1992). Traveling salesman games. *Mathematical Programming* 53, 199-211.

[100] Potters, J.A.M. and J.H. Reijnierse (1995). Γ-component additive games. *International Journal of Game Theory* 24, 49-56.

[101] Potters, J.A.M., J.H. Reijnierse, and M. Ansing (1996) Computing the nucleolus by solving a prolonged simplex algorithm. *Mathematics of Operations Research* 21, 757-768.

[102] Prim, R.C. (1957) Shortest connection networks and some generalizations. *Bell Systems Technical Journal* 36,1389-1401.

[103] Quinzii, M. (1984). Core and competitive equilibria with indivisibilities. *International Journal of Game Theory* 13, 41-60.

[104] Rafels, C. and N. Ybern (1994). Fixed points and convex cooperative games. Working Paper.

[105] Reijnierse, J.H., M. Maschler, J.A.M. Potters, and S.H. Tijs (1993). Simple flow games. *Games and Economic Behavior* 16, 238-260.

[106] Riker, W.H. (1962). The theory of political coalitions. Yale University Press, New Haven.

[107] Rochford, S.C. (1984). Symmetrically pairwise-bargained allocations in an assignment market. *Journal of Economic Theory* 34, 262-281.

[108] Roth, A.E. (1977). Utility functions for simple games. *Journal of Economic Theory* 16, 481-489.

[109] Roth, A.E. (1982). The economics of matching: Stability and incentives. *Mathematics of Operations Research* 7, 617-628.

[110] Roth, A.E. (1984). Stability and polarization of interests in job matching. *Econometrica* 52, 47-57.

[111] Roth, A.E. (1985). The college admission problem is not equivalent to the marriage problem. *Journal of Economic Theory* 36, 277-288.

[112] Roth, A.E. and M. Sotomayor (1990). Two-sided matching: A study in game-theoretic modeling and analysis. Cambridge University Press.

[113] Samet, D. and E. Zemel (1984). On the core and dual set of linear programming games. *Mathematics and Operations Research* 9, 309-316.

[114] Schmeidler, D. (1969). The nucleolus of a characteristic function game. *Siam Journal of Applied Mathematics* 17, 1163-1170.

[115] Shapley, L.S. (1953). A value for *n*-person games. In: *Contributions to the Theory of Games* II (H. Kuhn and A.W. Tucker eds.). Princeton University Press, Princeton, 307-317.

[116] Shapley, L.S. (1962). Simple games: An outline of the descriptive theory. *Behavioral Science* 7, 59-66.

[117] Shapley, L.S. (1967). On balanced sets and cores. *Naval Research Logistics Quarterly* 14, 453-460.

[118] Shapley, L.S. (1971). Cores of convex games. *International Journal of Game Theory* 1, 11-26.

[119] Shapley, L.S. and H. Scarf (1974). On cores and indivisibilities. *Journal of Mathematical Economics* 1, 23-37.

[120] Shapley, L.S. and M. Shubik (1954). A method for evaluating the distribution of power in a committee system. *American Political Science Review* 48, 787-792.

[121] Shapley, L.S. and M.Shubik (1972). The assignment game I: The core. *International Journal of Game Theory* 1, 111-130.

[122] Smith W. (1956). Various optimizers for single stage production. *Naval Research Logistics Quarterly* 3, 59-66.

[123] Sobolev, A.I. (1975). The characterization of optimality principles in cooperative games by functional equations (in Russian). *Mathematical Methods in the Social Sciences* 6, 150-165.

[124] Solymosi, T. and T.E.S. Raghavan (1994). An algorithm for finding the nucleolus of assignment games. *International Journal of Game Theory* 23, 119-143.

[125] Spinetto, R.D. (1971). Solution concepts of n-person cooperative games as points in the games space. Technical Report 138, Department of Operations Research, College of Engineering, Cornell University, Ithaca.

[126] Straffin, P.D. Jr. (1983). Power indices in politics. In: *Political and Related Models* (S.J. Brams, W.F. Lucas, and P.D. Straffin Jr. eds.). Springer-Verlag, New York.

[127] Tamir, A. (1980). On the core of cost allocation games defined on location problems. Department of Statistics, Tel-Aviv University.

[128] Tamir, A. (1983). A class of balanced matrices arising from location problems. *SIAM Journal on Algebraic and Discrete Methods* 4, 363-370.

[129] Tamir, A. (1989). On the core of a traveling salesman cost allocation game. *Operation Research Letters* 8, 31-34.

[130] Tamir, A. (1991). On the core of network synthesis games. *Mathematical Programming* 50, 13-135.

[131] Tijs, S.H. (1981). Bounds for the core and the τ-value. In: *Game Theory and Mathematical Economics* (O. Moeschlin and D. Pallaschke eds.). North-Holland Publishing Company, Amsterdam, 123-132.

[132] Tijs, S.H. (1987). An axiomatization of the τ-value. *Mathematical Social Sciences* 13, 177-181.

[133] Tijs, S.H. and T.S.H. Driessen (1986a). Extensions of solutions concepts by means of multiplicative ε-tax games. *Mathematical Social Sciences* 12, 9-20.

[134] Tijs, S.H. and T.S.H. Driessen (1986b). Game theory and cost allocation problems. *Management Science* 32, 1015-1028.

[135] Tijs, S.H. and F.A.S. Lipperts (1982). The hypercube and the core cover of n-person cooperative games. *Cahiers du Centre de Recherche Operationelle* 24, 27-37.

[136] Tijs, S.H., T.Parthasarathy, J.A.M. Potters and V. Rajendra Prasad (1984). Permutation games: Another class of totally balanced games. *OR*

Spektrum 6, 119-123.

[137] Tucker, A.W. (1960). On directed graphs and integer programs. Princeton University Press, Princeton.

[138] Wako, J. (1984). A note on the strong core of a market with indivisible goods. *Journal of Mathematical Economics* 13, 189-194.

[139] Wako, J. (1986). Strong core and competitive equilibria of an exchange market with indivisible goods. Tokyo Center of Game Theory, Tokyo Institute of Technology.

[140] Weber, R.J. (1978). Probabilistic values for games. Cowles Foundation Discussion Paper 417R, Yale University, New Haven.

[141] Weber, R.J. (1979). Subjectivity in the valuation of games. In: *Game Theory and Related Topics* (O. Moeschlin and D. Pallaschke eds.). North-Holland, Amsterdam, 129-136.

[142] Young, H.P. (1985a). Monotonic solutions of cooperative games. *International Journal of Game Theory* 14, 65-72.

[143] Young, H.P. (1985b) ed. Cost allocation: Methods, principles, applications. North-Holland, Amsterdam.

[144] Young, H.P. (1987). On dividing an amount according to individual claims or liabilities. *Mathematics of Operations Research* 12, 398-414.

INDEX

THEORY AND DECISION LIBRARY

SERIES C: GAME THEORY, MATHEMATICAL PROGRAMMING AND OPERATIONS RESEARCH
Editor: S.H. Tijs, *University of Tilburg, The Netherlands*

KLUWER ACADEMIC PUBLISHERS – DORDRECHT / BOSTON / LONDON